The Best of *Ruminations* 2001–2007

Goat Health Care

Cheryl K. Smith

Illustrated by Myra Klote

karmadillo Press • Cheshire, Oregon

karmadillo Press
22705 Highway 36
Cheshire OR 97419
USA
(541) 998-6081
www.karmadillo.com

Author and Editor: Cheryl K. Smith
Illustrations: Myra Klote
Proofreaders: Wanda J. Walker and Kate E. Cole
Cover Design: Cathy Guy

Copyright 2009 © Cheryl K. Smith

Photographs, pages 90, 91, 99, 199, copyright 2007 © Cheryle Moore-Smith

Notice of Rights
All rights reserved. No part of this book may be reproduced or transmitted in any form or by any means, electronic or mechanical, including photocopying, recording or any information or retrieval system, without written permission from the publisher.

ISBN: 978-0-615-24484-6
Printed and bound in the United States of America

For more information:

www.goathealthcare.com

Dedication

To my first goat, a Nigerian Dwarf I got in 1998 named ARMCH OTR Magic's Hijinx *D, better known as Jinx.

She got me started in the goat world and kept me motivated as she won show after show. She eventually made the Top Ten in Nigerian Dwarf milk production in the AGS Milk Test program. Jinx spent too little time in this world, dying at the age of seven. Her contribution to my knowledge of goats was immense, and now I hope to pass that on to you.

Disclaimer

The information contained in this book is not intended to replace professional veterinary advice or manufacturer's prescribing information. The anecdotal information, experiences and thoughts are those of the authors and are not meant to represent the management practices or thinking of goat breeders in general, or of veterinary professionals.

Individuals and their veterinarians must make responsible decisions concerning treatments and drug safety or effectiveness. The extralabel use of any product in a food-producing animal is illegal without a prescription from a veterinarian. Some drug information may change without notice and readers are advised to consult their veterinarian(s), the Food Animal Residue Avoidance Databank (FARAD), and/or manufacturer's insert prior to giving a medication to their goat(s).

Table of Contents

Understanding Your Goat..1
    Troubleshooting Goat Health Problems...............................2
    Goat Health Record...6
    The Ruminant Stomach...7
    Goat Teeth..9
    Goat Sociology...10

Routine Care..12
    Fencing and Shelter...13
    Feeding..14
    The Stanchion...16
    Rotational Grazing..18
    The Ideal Pasture..20
    Supplements...21
    Hoof Trimming...27
    FAMACHA..30
    Creating "Super Worms"..32
    Parasite Control..33
    Giving Injections..37
    Drawing Blood..40
    Vaccinations..42
    Cold Weather Goat Care...43

Breeding, Pregnancy and Kidding...46
    Pre-breeding..47
    Inbreeding—Good or Bad?...49
    Why AI?...51
    Hypocalcemia: How to Recognize, Treat and Prevent It...53
    Ketosis: What Is It and How Does It Happen?.................61
    Abortion in Goats of North America................................63
    How to "Read" Ligaments..68
    Kidding for Beginners..69
    A Difficult Kidding...74
    Other Kid Positions..77
    A Difficult Kidding: Part II...78
    Acquired Birth Defects in Kids..79

Kid Care...81
    Post-Kidding...81
    Tube Feeding a Weak Kid...83
    Floppy Kid Syndrome..85
    Correcting Hyperflexed Legs in a Newborn Kid.............86

Navel Ill...87
Feeding Kids..87
Why Disbud?..88
Disbudding—A Learn-as-you-go Adventure.......................................89
Castrating Your Buckling..93
Tattooing and Microchipping...97
A Word About Milk Goiter...100
Coccidiosis..101

Health Issues...105

Bloat..105
Brucellosis..106
Caprine Arthritis Encephalitis Virus (CAEV)....................................107
Caseous Lymphadenitis (CLA)..113
Caprine Chlamydiosis..115
Copper Deficiency in Dairy Goats...118
Tyzzer's Disease and Copper Deficiency...124
Coughing Caprines...127
Enterotoxemia..131
Foot Rot...133
Johne's Disease...134
Those Pesky Lice!...136
Listeriosis...138
Mastitis...139
Gastrointestinal Parasites..142
Pinkeye...145
Pneumonia...147
Polioencephalomalacia (Goat Polio)...149
Poisoning..152
Prussic Acid Poisoning..154
Q Fever...156
Rabies..157
Ringworm..159
Scrapie..160
Soremouth..161
Sunburn..162
Tetanus...163
Preventing Urinary Calculi...164

Medications...167

Injectable Antibiotics...172
Injectable Vitamins and Minerals...178
Miscellaneous Injectables..180
IV/SQ Solutions..186
Oral Antibiotics/Antibacterials..188
Miscellaneous Oral Products..190

    Oral Pain and Anti-inflammatory Medications..................................192
    Other Miscellaneous Products..193
    Mastitis Detection and Treatment..195
    Drugs for Intramammary Use in Non-lactating Cows........................196
    Anthelmintics (Dewormers)...197
    Anticoccidial Products...198

Natural Care and Home Remedies..199
    Home Remedies..200
    Herbal Remedies...205
    Fighting Flies in the Goat Barn...208
    Alternatives: Herbal Remedies..209
    Alternatives: Teeth..210
    Alternatives: Breeding..212
    Alternatives: Pre-kidding...213
    Alternatives: Parasites..215
    Goat Massage 101..217

Letting Go..220
    Euthanasia..223
    Disposal of the Body...224
    Bereavement..227

Resources...228
Index..240

Acknowledgements

This book would not have been possible without the help of a number of people.

Joy Lewis of Over the Rainbow Farm, sold me my first two goats and was my first mentor in caring for them.

Stacy Morris of Flyaway Farms was available to answer any goat question I had, at any time, as well as writing for *Ruminations* magazine.

My good friend Rubbin proofread many of the original articles in this book and was always willing to work on a short time line, when necessary.

Wanda Walker, with her eagle eye on layout and proofreading, helped me with the book's appearance.

Cathy Guy created the perfect cover design.

Jan Tritten, who gave me the flexibility to complete and publish this book and the opportunity to learn a lot about self-publishing.

Myra Klote, who was always willing to do another goat drawing, made the book more practical and pleasing to the eye.

All of the other contributors and subscribers to *Ruminations,* without whom this book would not have been possible, especially Donna Geiser, Joyce Lazzaro, Cheryle Moore-Smith and Sue Reith.

Introduction

I read somewhere that the average goat keeper has goats for only five years. This is the time in which reality sets in and the goat owner is faced with at least one serious health care issue. It may be a chronic cough that develops when the weather is unpredictable, scours in a kid, or a problem with kidding.

Let's face it, with goats still considered a minor species, it's challenging to find a veterinarian who has expertise in goats or whose practice provides services to them. We most often have to provide our own goat health care, or have to acquire the information on our own so that we can educate the veterinarian. Another stumbling block is that many medications haven't been approved for use in goats, so we have had to learn what works through trial and error, as well as finding a willing veterinarian to prescribe them.

The first step in keeping goats healthy is to start with disease-free, thriving goats. The ideal foundation for a new herd is kids that were dam-raised (nursed by their mothers) on does that are disease-free. This allows the kids to obtain the ideal goat food—natural colostrum and milk, which are what nature intended for a healthy immune system. (This is not to say that kids raised on milk replacer or pasteurized milk cannot do well.)

Goat buyers should ask whether the goats being purchased have been tested for any diseases and/or whether their dam has, as well. Ask about overall herd health, how the goats' health care is managed and what kinds of veterinary problems they have had. Find out the breeder's philosophy on goat care. Some use only herbal care and have had success with that, others use veterinary medications, and many use some of each. Find one whose goats look good and whose herd health management is compatible with your views. Ask lots of questions!

This book is a compilation of health care articles from seven years of *Ruminations* magazine, with additional information included to make it more helpful. It comes out of my ten years of experience breeding and raising goats and seven years as a goat magazine publisher. Throughout this time I have had to take care of my own goats (often with no veterinarian assistance), learn from experiences that other goat keepers have shared with me and advise newbies on what to do when they face similar problems in their herds.

Goat Health Care is not meant to be a comprehensive book of goat medicine; I will leave that to the professionals. But it is the only book on the market geared to the small goat herd that contains a comprehensive list of veterinary medicines that are used for goats—both approved and extralabel, as well as herbal goat care and dealing with end of life issues.

I also have included a resource section that provides further references and information that isn't contained in this book. With the Internet at our fingertips, we now have access to a huge number of resources.

One of my intentions in writing this book was to help new goat owners avoid some of the mistakes I made and the heartaches I felt when something went wrong with a goat. Once you learn the ins and outs of goat keeping, please pass it on to others. We are all in this together.

Finally, I can't stress enough the importance of developing a relationship with a veterinarian. By doing so you will achieve two goals: obtaining the necessary medication for your goats and educating each other so you can do a better job as a goat keeper and other goat owners and their animals can benefit from the expertise.

Enjoy your goats!

Cheryl K. Smith
Low Pass, Oregon
December 5, 2008

Understanding Your Goat

"All goats are mischievous thieves, gate-crashers, and trespassers. Also they possess individual character, intelligence, and capacity for affection which can only be matched by the dog. Having once become acquainted with them I would as soon farm without a dog as without a goat."

~ David Mackenzie, *Farmer in the Western Isles* (1954)

Goats are believed to have been domesticated more than 10,000 years ago. Small herds were kept for milk and meat, as well as for clothing and building materials. Their dung was even used as a fuel source, as well as for garden compost. Most of us these days don't use the full array of products that they can supply, but we have still maintained our relationship with these fascinating animals.

Goats have been referred to as the "poor man's cow," but anyone who has raised goats knows that they provide a much richer experience than any cow—they are a lot smarter, more efficient and much more entertaining than a cow. Each one has a unique personality, cry, way of moving, and look, whether it be an Oberhasli, a Nubian, a Boer, a Nigerian Dwarf or any other breed.

Although they originated in a desert region, goats can live and thrive in a variety of climates, as long as their basic needs are met. These include clean bedding and shelter, a balanced diet and routine care. But the real key to goat health is observing and understanding your goat. If you don't know what is normal for that goat, you can miss some subtle cues that a problem exists or is evolving.

Not only should a goat be observed for its individual behavior, but for its behavior in and among the herd. How these social creatures interact can tell you a lot about how they are feeling and, consequently, how other goats are feeling about them.

Troubleshooting Goat Health Problems

Basic Physiological Data

Temperature (rectal)........101.5–105.0° F.

Pulse...................................70–80 beats/minute

Respirations.......................12–15 breaths/minute (faster for kids)

Ruminations......................1–1.5/minute

Puberty..............................2–12 months (Nigerians may breed at 2 months)

Estrus (Length of Heat)....12–36 hours (avg. 18 hours) every 21 days

Gestation............................145–155 days

Birth weight.......................Depends on breed and number of kids (.5–9.0+ lb)

Temperature

The rectal temperature of goats can vary tremendously depending on factors such as the length of hair coat, ambient temperature and level of excitement. Before deciding whether or not a particular animal has a fever or subnormal temperature, take rectal temperatures of at least two other animals in the herd.

> **To take a goat's temperature:** Shake the thermometer down, lubricate with KY jelly, insert it in the goat's rectum and leave it for 3 minutes.
>
> Compare the results to the goat's normal temperature (if known) or that of other goats in the herd.

On a hot summer day, a doe with acute mastitis may have a rectal temperature of 106.5° F. while "normal" animals housed with her may have 104.0° F. rectal temperature. In cold weather, a doe with acute mastitis may have a rectal temperature of 105.5° F., when "normal" animals housed with her have temperatures of 102.5° F.

A rectal temperature as high as 106.5° F. can be common in an apparently normal animal that has been chased to catch her in hot weather. The only way to be reasonably sure that a particular animal's temperature is abnormal is to compare it with others at the same time or to take it on several occasions when she is well.

A thermometer is one of the most important tools in a goat keeper's medical toolkit. Digital thermometers are now reasonably priced, and because they are easier to read and faster than regular thermometers, they are a good investment.

You the Caretaker

Observe your animals and learn the habits and personality of each of them. A change in habits, personality or routine will signal you that a problem is beginning and you can begin proper care hours or even days before a condition becomes serious. If a goat is acting "different" or "off," investigate at once—do not wait until later, or the next morning. Examine the goat, make sure that it is eating, ruminating, urinating and has normal poop, take temperatures, etc. If you have any doubt, call a goat friend or your veterinarian (now, no matter the time of day or night), and talk it over with them. Catch the problem early.

> **Tip:** If a goat is acting "different" or "off," investigate at once—do not wait until later…

Trust your judgment; you know more about your animals than you think you do and you can do more in emergency situations (e.g., problem kiddings) than you'll ever imagine you can. Keep a cool head and trust your "gut feelings"; they'll serve you well.

You and Your Veterinarian

No matter how well you manage your herd, or how many experienced goat friends you have, you will still need the services of a veterinarian. The money for proper medical care of your animals is more than well-spent and will be returned to you in the long run.

Try to locate and establish a working relationship with a veterinarian before you have a problem. Find a vet you are comfortable working with; work with that vet and he/she will work with you. Don't wait until your animal is near death to call the veterinarian (and then blame or complain when he/she is unable to save the animal). Follow your veterinarian's instructions and suggestions. Don't be afraid to offer your thoughts and observations to your veterinarian: The two of you are a team, it's not one against the other. If you can't establish a good team relationship with your veterinarian, find another vet.

If no veterinarians with goat experience are available in your area, work alone or with other goat breeders to educate the local veterinarian(s) about goats. Send clippings of new information on goats; share health information in goat publications or buy him or her a subscription to the *American Association of Small Ruminant Practitioners* (AASRP). (See Resources)

Evaluating Your Goats

The good goat keeper will observe his or her animals regularly to make sure that they are healthy. Ideally, when things are not right, you will make a note of any problems. A variety of forms have been designed for this purpose. You may be able to find one on the Internet, at a goat conference or through a friend, or you may want to design your own based on what you learn in this book and through experience with your goats. I designed a simple form (see page 6) for recording medical information on my goats. You can use this format, or design one of your own.

> **Trick:** If you need to examine a goat that is resistant, raise its head and it can't get away.

To determine whether your goats are healthy, **observe** each of them at least once or twice a day. Look for the following:

1. Is the goat ruminating (chewing its cud)? As a general rule, at least 1/3 of the goats should be chewing their cuds at any one time.
2. Is it walking normally? If not, why not? Is there a foot problem? Are the knees all right? Does the goat have an injury?
3. Is it acting normal (for that goat)? Is a normally engaging goat standing off from the herd? Is it lying down, rather than standing up? Can it see? Is it exhibiting an abnormal body posture?
4. Is the goat grinding its teeth, pressing its head or showing other signs of pain?
5. Is it breathing normally? Count the respirations per minute.
6. Does the goat have normal pelleted poop? Is it urinating normally?

If you see a goat that needs further investigation, **examine** that goat. Depending on what you notice about the goat, do the following, noting the results on the goat's health record:

1. Take the goat's temperature.
2. Place your palm or your ear on the goat's left flank and feel, or listen, for rumen movements and gurgling sounds. Watch for signs of pain and note whether the goat's rumen feels bloated.
3. Determine the goat's heart rate by placing your fingers on the goat's lower rib cage and counting the heart beats for 30 seconds. Double that amount for beats per minute.
4. Check the gums and eyelids to see the color of mucous membranes. If they are pale, the goat may be anemic, which is most often a sign of a parasite overload.
5. Check the goat's body for swelling, injuries or other problems. Move your hand toward the goat's eye to check for blindness.

Goat Health Care

6. Listen for sounds that may indicate a problem, such as wheezing or coughing. Listen to the lungs and heart on the right side with a stethoscope, if you have one.

7. Evaluate the eyes and nose, noting whether they are runny, mucousy or crusty.

8. Look at the goat's ears to see if they are scratched or held oddly. Scratching can indicate mites. A droopy ear can be a sign of an ear infection.

9. Check the udder in lactating does, noting whether it is hot, has lumps or has other abnormalities.

10. Check for diarrhea and, if necessary, collect feces for submission to a lab or examination under your microscope.

11. Check the goat for dehydration by pinching the skin on the neck in front of the shoulder, using your thumb and forefinger. Note whether the skin snaps back to its normal position quickly or responds slowly and remains "tented." A slow return to normal can indicate that the goat is dehydrated.

The data you have collected from this simple evaluation will help determine what is wrong with the goat. After you have raised and studied goats for a while, this information will enable you to narrow down the possibilities of what might be wrong with your goat. If that isn't the case, you may be able to find the answer in this book or another one, by sharing the symptoms with other goat owners or a veterinarian, or by searching the Internet. (See Resources)

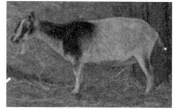

Above: Note the posture of this sick goat.

In some cases, you may not be able to determine exactly what is wrong with the goat, e.g., it has a temperature of 105.2 F. (when its normal temp is 104) and it is coughing, it has no diarrhea, and no other signs of infection can be found. In this case, consult a veterinarian, or if you don't have access to a veterinarian communicate with other goat people, give the goat some probiotics, some Bo-Se or vitamin B-complex, and consider starting an antibiotic such as Naxcel (Rx) or Bio-Mycin (OTC). Then continue to observe the goat.

Trick: To move a goat that is down and can't get up, roll it onto a carpet strip and pull the carpet to the desired location.

Goat Health Record

Name of Goat_____ Breed_____

Date of Birth_____ Tattoo/Microchip #_____

Sire_____ Dam_____

Health Information

Date	Treatment/Injection	Comments

Kidding Information

Breeding Date: Due Date:
Bred to:
Kidding Date: # of Kids/Sex:
Outcome/Comments: e.g., not settled, kids weak or died

Notes:

The Ruminant Stomach

Knowing how your goat's stomach works will go a long way in helping you maintain the goat's health. Goats are ruminants, which means they are members of a subgroup of mammals that ruminate, or chew their cuds.

Ruminants, like humans, have only one stomach. You may have heard that they have more, but it is actually one stomach divided into four parts—three of which are considered forestomachs. The fourth part is a true stomach. Learning how this system operates is essential to good herd management.

A ruminant digestive system consists of a rumen, a reticulum, an omasum, and an abomasum. The first three of these compartments are responsible for grinding and digesting hay and some grain, with the help of bacteria. The last compartment, the abomasum, is similar to the human stomach and is a digestive organ for most proteins, fats and carbohydrates.

Newborn Kid's Digestion

The newborn kid initially functions as a simple-stomached animal. When the kid nurses, a band of muscle tissue called the esophageal groove closes so that milk bypasses the first three stomachs and goes directly to the abomasum for digestion. If this groove doesn't close during early feeding, milk will go into the immature rumen and may curdle and cause symptoms of colic in the kid. That is why it's important to hold kids in a natural nursing position if they are being bottle-fed.

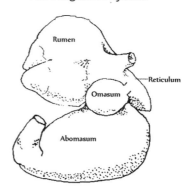

Kid's Digestive System

Maturing kids will start to nibble on hay, grain (and everything else), and normally will have a functional rumen around one month of age. This maturation of the rumen comes about through the introduction of microorganisms that are necessary to its function. Dam-raised kids' rumens develop more quickly than those of bottle-fed kids, in part because they are learning from the adult goats how and what to eat, but probably also because they are more likely to have exposure to the microorganisms found in the poop of other older herd members.

Rumen

Each of the four stomach compartments has a particular purpose and function. The rumen is the largest chamber (1–2 gallon capacity) and has no digestive enzymes. It is a large fermentation vat populated by microorganisms that turn nondigestible cellulose into protein that can be used by the body. Rumen bacteria break down some of the roughage, such as hay, and then it is regurgitated, chewed by the goat as a cud and then swallowed. This process is repeated, in what is known as rumination. Because methane is produced continuously as a by-product of the bacterial action, strong-smelling belches are a sign of healthy rumen. Anything that harms or destroys rumen microorganisms can effectively halt the digestive process.

> **Tip:** Where kids are raised separately from the rest of the herd, throw adult goat berries into that area to provide the kids with the necessary bacteria for development of rumen function.

When a ruminant is unable to belch, the condition is known as bloat. Bloat can be life-threatening. For mild cases, the first line of treatment is milk of magnesia.

Reticulum and Omasum

Once the roughage is broken down, it goes through the reticulum to the omasum to be digested by normal body enzymes. The reticulum lies in front of and below the rumen, near the liver. It has a honeycomb-like lining and serves as a catch chamber for heavy articles in feed—more of a problem in cattle.

> **Just for Fun:** To see what the different parts of a ruminant's stomach lining look like, go to:
> http://education.vetmed.vt.edu/curriculum/vm8054/Labs/Lab21/lab21.htm

The omasum is divided by long folds of tissue that help to decrease the size of food particles coming from the rumen and help remove excess fluid.

Abomasum

The abomasum is considered the "true stomach," and is the only compartment that produces digestive enzymes. It acts on food that has been partially broken down by the forestomachs, just as our stomachs acts on food that we eat. This is where primary digestion of grain and milk occur; they don't need rumen bacteria for digestion.

Goat Teeth

Goats have lower teeth in the front of the mouth, but the upper front of the mouth has only a hard pad, often called the "dental pad." Kids have six of the lower incisors at birth. Within a month or so, their teeth are complete and consist of 20 "milk teeth." These include two more incisors in the front of the lower jaw (making eight total), and 12 molars, three on each side in each jaw. I advise children to keep their hands out of a goat's mouth because these molars—top and bottom—can really hurt! Goats use these teeth for cud-chewing.

Baby goats' initial teeth erupt at around 98–105 days gestation. They lose these baby teeth, just like other mammals. You can generally determine the age of a goat by looking at the eight teeth in the goat's lower front jaw. This is called "toothing" a goat. Some natural variances occur in the loss of teeth and growth of new ones, and other factors, such as diet and health, influence growth of teeth; so this information provides only an estimate.

First year (kid) Baby teeth are small and sharp. These gradually fall out and are replaced by permanent teeth.

Second Year (yearling) The two middle front teeth fall out when the kid is about 12 months old. Two larger permanent teeth replace them.

Third Year (2–3-year-old) The teeth next to the middle (now permanent) teeth fall out. Two new, larger permanent teeth replace them when the kid is about 24 months old.

Fourth Year (4-year-old) The next two of the eight fall out and new permanent teeth grow in.

Fifth Year (over 4 years old) All eight front teeth are now in.

After five years, wear on the teeth must be used to gauge the goat's age. This will vary immensely, based on diet. Also, like people, some goats' teeth loosen over time and fall out as the goat ages.

Tip: Check the teeth of an older goat that is off feed or losing weight. It may need to be switched to "blender soup," such as pelleted senior horse feed soaked in hot water, to obtain enough nutrition.

Goat Sociology
by Cheryl K. Smith

You can't have a goat herd for long without becoming something of an amateur goat sociologist. Sociology is the study of groups, and goats are an excellent group to study. (Actually, the correct term for study of animal behavior is ethology.)

> **How many goats should you get?** This may seem like a strange question, unless you know goats. They are herd animals, so having just one goat is usually a bad idea. It will get lonely and bored, unless paired with another animal like a horse or a sheep. Goats are not generally good matches with dogs, in terms of companionship. So assume that you are going to purchase two goats.
>
> From *Ruminations* #42

Learning about how goats interact with each other is as important as learning how to feed and care for them as individuals, whether they are pets or producers.

In their natural state, goats tend to establish a dominance pattern within their herd. Age and sex are determining factors in caprine dominance patterns. (Horns can also play an important role, but since most dairy goats are disbudded or hornless, this article focuses on the other factors.) When does and bucks are kept separate, age becomes the most important factor in dominance, although as any goat owner can attest, personality plays a definitive role.

The herd has a herd queen, who leads the other goats to the best grazing areas (although she may try to keep the best for herself, her daughters and her chosen friends). My observations have indicated that this is not always the case, but when she does go, everyone follows. Removing the herd queen can confuse the other goats.

In the wild, goat herds also have a top buck, who will take up the rear of the herd to protect against predators.

The dominance pattern among goats is established with the original herd. It is likely to continue until the lead doe or buck dies, or in some cases, gets old and/or infirm. A discussion on one of the e-mail goat lists dealt with a herd queen being challenged for her position shortly after she and the challenger gave birth. After several separations, the two gave up their battle, with the original herd queen giving up her position to the challenger. I have witnessed this type of challenging behavior after kidding, especially if the herd queen has been removed from the herd and placed in a kidding pen for a period of time. While she is out of the herd, the next goats in line have been staking out their territory, only to discover (usually) that the situation was temporary.

Goats establish dominance through head-butting and other aggressive acts, such as biting. The dominant goat will get to eat first, drink first, do everything first!

In the wild, or when they are allowed to run together, the top buck will dominate the herd, with the herd queen second. When separated, the herd queen will dominate the group of does and kids. The herd queen will always be dominant over her doe kids, even after many years.

> **Tip:** Because they are prey animals, goats will "cover up pain-related behavior when people are watching. They do this in the wild to avoid being eaten by predators."
>
> Temple Grandin
> from *Animals in Translation*

I've observed in my herd that the doe kids of my herd queen automatically move up near the top of the herd in dominance. They are obviously used to special treatment as, from their birth, their mother would run off any goat that got too close (other than her older daughters).

According to a fact sheet written by E.A.B. Oltenacu and Tatiana Stanton for the New York State 4-H Dairy Goat Project, in the wild, when the herd queen finds a poisonous or inedible plant, she will sniff it and then snort and show obvious dislike for it. All the goats in the herd will then smell the plant, one at a time, using the scent to identify that plant. Then after all the goats have memorized the plant's scent, the top buck will trample the plant.

Some natural goat behaviors may not occur in domesticated herds, because of different conditions that affect how they act. These include all the ways in which humans interact with their goats. For example, when humans feed the goats they become associated with the herd queen and may have problems getting the goats out to browse. My goats are so conditioned by my feeding, that when I drive up in my car, or walk from the house to the barn, they all run in from the pasture.

Goat watching is fun and always brings the new and unexpected, as well as the old and expected. Now take some time to go out and observe your goats' behaviors.

From *Ruminations* #29

Just for fun: Watch your goats for herd behavior. Can you see some of the behaviors discussed? What other kinds of behaviors do your goats exhibit?

Routine Care

Igloo Shelter

Three-sided Shelter

Fencing and Shelter

Fencing and shelter are important to keeping goats healthy. Fencing not only keeps goats in a safe area, it also keeps predators *out*. Before getting goats, see what other goat keepers do, read up on fencing options and find the best one for your situation. Factors to consider include the size of the goat (they are notorious for escaping, so the fencing must be high enough), your location (rural or suburban), the surrounding area and your finances, among others.

Cattle panel, which comes in 16-foot lengths and is five feet high, is one of the easiest ways to fence. Smaller kids can get through it, though, so keep this in mind. Some people have had good luck with electric fencing, and field fencing keeps even the smallest kid in. (See Resources for fencing ideas.)

NEVER tether a goat. Tethered goats are bait for predators, such as dogs, coyotes and cougars. Because goats are browsers, they are likely to get their tether tangled up by moving around. I also have known of goats to hang themselves when they were tethered near trees. Tethering a goat is just asking for trouble and, for me, is a deal-breaker when it comes to selling to a new owner.

Tip: When designing goat housing, think about goat behavior. For example, sometimes a goat will block other goats from coming in. To prevent this problem, include two exit/entrance doors in the plan.

Goats need housing to keep them out of wind, rain and cold, minimally. Depending on their location and whether or not they are protected by a guardian animal, they also may need housing where they can be closed up securely.

Housing can range from a dog igloo to a garage or shed, to a full-fledged barn. The health-related goal of the housing should be to keep goats from weather extremes and drafts, as well as to give them a secure place to sleep.

Also consider the type of floor (everyone has a preference, and I prefer dirt), bedding to be used (think about cost and the ease of mucking), and sleeping and feeding setups. Goats like to be up on something, so consider a "sleeping shelf" that gets them off the ground. Also, if you will be milking, you will need a separate area for that, and also for storing feed.

See what others are doing and what options are available, and then evaluate specific needs before choosing housing. Talk to other goat owners, look at web sites and read up on the subject prior to making a decision.

Feeding

Unless you have a large and diverse acreage, the biggest expense in keeping goats (once you have purchased all of the necessary equipment, fencing and housing) is likely to be feed. This is a worthwhile expense, because proper nutrition will keep your goats healthier and more productive, saving you on veterinarian bills and heartbreak down the road. Despite what some people believe, though, goats need more than a pasture full of grass.

Water. Before I got goats, a neighbor with a pet buck told me that goats don't need supplemental water because all of their needs are met by the water in the plants they eat. This is a myth. While some forages may consist of 70–80% water, goats still need a source of fresh clean water, both for hydration and to prevent disease. This is even more important for dairy goats that are producing milk.

> **Tip:** Goats prefer warm, flavored water over cold, plain water.

Body Condition. Goats should not be fat, nor should they be overly thin. Lactating does, as well as growing kids, have greater nutritional requirements than dry does, bucks or wethers. Body condition is probably *the* best indicator of whether a goat's nutritional requirements are being met. Having said that, even thin goats have been found to have internal fat upon necropsy.

> **Tip:** Avoid hay that has been rained on or is moldy. Hay that has been rained on may not be as nutritious as hay that is properly cured. Moldy hay can make a goat sick or kill it.

Beyond seeing how a goat looks to determine whether it has proper nutrition, few hard and fast rules exist for feeding goats because nearly every goat keeper's situation is different. Some goats have very little pasture; others have many fenced acres full of a variety of forage. Weather conditions vary. Goats dislike rain, which will affect even their willingness to browse during certain times of year in areas such as the Pacific Northwest. Those in the desert may find little in the way of browse that they want to eat.

Feed. Depending on the size of your herd, feeding alternatives will vary. For instance, a goat keeper with 100 goats will most likely find that formulating a custom feed mix is the most cost-effective and healthful approach. On the other hand, one with only two goats could end up with moldy or stale feed if each ingredient were bought in bulk for mixing. That person may be better off with a commercial goat mix.

Regardless of what feed is used, make sure to store it in a dry, secure area to keep out mold, insects and rodents. Some people use plastic garbage cans; I prefer metal ones secured by a bungee cord.

Generally, dairy goats eat about 2–4% of their body weight in dry feed per day. Smaller goats need a higher proportion to maintain their weight than larger goats. Of course, the amount of energy they expend will also affect the amount of food needed, as will climate. So a very active goat or one that lives in a cold climate needs more food. In addition, pregnancy and lactation require more energy, and goats that are still growing (kids) do, too.

> **Tip**: There is no reason to give a pet goat any grain.
> Veterinarian Mary Smith
> Cornell University

Goat feeds can be divided into two basic groups, roughages and concentrates. Roughages are high in fiber and consist mainly of forages that come from the green parts of plants. These include hay, grass and other plants. The fiber in roughages is essential to the goat's digestion, adds bulk to the diet, and also increases butterfat content of milk. On the other hand, roughages are low in energy.

Roughages can include dry forages such as hay or straw, which are cut and cured. This makes them ideal for storage and winter feeding. Roughages can also include the various plants that goats browse on in the pasture, such as grass, blackberry bushes or various shrubs.

Concentrates are low in fiber and high in energy or protein. They include corn, oats, barley and soybeans, among others. Some concentrates provide more energy, while others have more protein; some are high in both protein and energy.

I recommend reading about optimal goat feeding, talking to other goat keepers, working with the local extension office to develop a feeding program, and/or searching the Internet for more information on feeding and balancing rations. (See the Resources for some good books on this topic.)

Tips for Feeding Dairy Goats

- Feed kids enough energy for growth, and feed mature animals enough energy to maintain their body weight.
- Provide appropriate protein, minerals and vitamins.
- Give does extra feed during pregnancy and lactation.
- Don't overfeed grain in relation to hay or forage.
- Gradually introduce goats to fresh, green pasture and cuttings.
- Don't feed poisonous plant clippings, such as yew, rhododendron, azalea, or cherry, apricot or peach branches.

The Stanchion

Whether you have two goats or 20, a stanchion is one of the best pieces of equipment you can buy. A stanchion is a device that locks the head of a goat in place so that you can milk it or perform routine care such as hoof trimming, injections or clipping. Ideally it is attached to a stand so that the goat is elevated, making such work easier, and has an attached container for feeding grain or other food to distract the goat.

Often called a milk stand, because it is used while milking goats, this stand with stanchion can be made from wood, PVC or metal. Milk stands can be purchased from a goat supply company or used (on rare occasion), or they can be homemade. Free-standing stanchions can also be purchased and used to make a milk stand. If you are handy, making one is the best way to go, and directions are available on the Internet.

Goat keepers have made milk stands in a variety of ways. They can be made to break down and be light enough to make travel to a goat show or fair easy. They can incorporate storage, they can be different heights or even adjustable for the various sizes of goats.

Goats soon learn to eagerly jump onto the stand—not because they like being there, but because they want the food that is being offered. Even recalcitrant goats can be trained to get on the stand if taken to it and allowed to eat quietly a few times.

Metal milk stand with holder for grain bowl. A paper towel holder has been installed above and bungee-secured garbage can for grain is placed near the head, for convenience.

Say No to Cottonseed in Goat Feed

A short article was published on the deaths of four goat guardian dogs who died of heart problems over the course of three years, usually in the winter. (Pannill et al. April 2006) The culprit was cottonseed in the goat feed.

Cottonseed contains a high concentration of gossypol, which is a toxin that affects primarily the heart and liver. Ruminants can tolerate higher levels of gossypol than animals with only one stomach because it binds to proteins in the rumen. An animal with only one stomach, such as a human, a dog or a pig, is much more susceptible to the poison. However, even young lambs and calves have been found to have toxic reactions to gossypol; that can probably be said about young kids, as well. This is due to the fact that their rumens are not yet fully developed and, in many cases, occurs when they are on free choice feed.

Cottonseed and cottonseed meal are now used as additives in livestock feed. This is important to know if you have a livestock guardian dog that develops depression, difficulty breathing, appears to be having a heart attack and ultimately dies, and you know that the dog consumes goat feed. Gossypol toxicity also may be a consideration if you have a number of kids seeming to die of overeating disease, or developing chronic, difficult breathing, unthriftiness, failure to respond to antibiotics and going off feed. These can be signs of such poisoning.

No treatment currently exists for gossypol poisoning, although its course can be reversed with removal of the cottonseed in feed for a period of time.

The writers estimated the amount of goat food that a dog would need to have consumed to cause their fatal heart damage was only 3/4 cup daily for a few weeks. (Kid feed contained .01% and the does' was .06%.)

You can do at least two things to avoid situations such as this: 1.) Read the feed tag and make sure you know what you are feeding your animals/aren't buying feed with cottonseed or cottonseed meal; and 2.) Keep your guardian dog away from the goat feed. To read more, see the Oklahoma Cooperative Extension web site at www.osextra.com and look for the fact sheet "Gossypol Toxicity in Livestock."

From "Health and Science News," *Ruminations* #53

Rotational Grazing

by Nate France

A pasture has the potential to provide all the food a ruminant animal needs year-round. With feed costs as high as they are, using your our own land to grow feed would seem highly desirable. Many pastures, however, are little more than weedy exercise lots, heavily subsidized with brought-in feed. In order to control weeds and grow quantities of good ruminant food we need to encourage forage growth and soil fertility.

Pasture weeds, such as tansy ragwort (which is lethal to horses and cattle, but less so to goats) and Canada thistle, generally increase in number when animals have constant access to a pasture. Although you can kill these weeds with an herbicide or perhaps with repeated mowing, they will return unless you adopt management that favors grass over weeds.

Grass needs to rest after grazing or cutting in order to build up its strength again. Since most livestock prefer grass to the weeds in the pasture, they will graze the grass repeatedly and leave the weeds alone when given a choice. Grass cannot compete under these circumstances.

In the wild, grazing animals will generally not stay in the same place for very long because of predator pressure and a "the grass is always greener over there" penchant. The plants are grazed briefly and trampled, and then allowed a long rest period. These conditions give the grass a competitive advantage because it can regrow more rapidly and fully than most weeds, thus more effectively competing for water, nutrients, and light. Competition with vigorous grass ultimately reduces pasture weed populations.

How do you mimic this natural system when you must fence in your animals?

You can divide your pastures into three or more paddocks, allowing the animals into only one paddock at a time. This is known as rotational grazing. It allows the grass in the other paddocks to rest and regrow without being repeatedly bitten off. Some people use this basic rotation principle to get very high grass yields per acre by making many small paddocks with temporary electric fence, concentrating the animals in one paddock at a time and moving their animals every 24 hours or less frequently.

The more animals in a paddock, the more evenly they will eat down the forage, and the more paddocks you have, the longer the rest period will be for the grass. This is known as management intensive grazing (MIG) and several good books are out on the subject. Managing your grazing this way can result in higher total yields of both grass and animal products for your acreage.

How do you know when to move your animals?

In order for our common pasture grasses to regrow vigorously, a 3–4 inch stubble height needs to be left at the end of each grazing period. This is because most of the energy stores for these grasses are present in the bottom 3–4 inches of the stems. When the plant regrows it uses these energy stores to make new leaves. If the grass is bitten off below this height, regrowth will be slow, and if it is bitten off repeatedly below this height, it will die. Move the animals out of the paddock before they get a chance to take the plants down below three or four inches.

When should you move your animals into a paddock?

Let the grass grow to 6–8 inches before allowing the animals to graze it again. If growth gets much taller than this and especially when grass begins to go to seed, it redirects energy into reproduction and will not regrow as quickly when cut again. When the grass is growing so fast that you can't keep the height down to 6–8 inches, you can either temporarily add more animals or cut some paddocks for hay. Be sure to mow hay at the same 3–4 inch stubble height for quick regrowth.

Tip: If you have limited space, consider temporary fencing as an alternative to separate paddocks for rotational grazing.

At times the soil will be too soggy to support livestock without compacting the soil and reducing its ability to grow forage. At other times, such as during a drought, the forage will be growing so slowly that the grass in the next paddock has not yet reached 6–8 inches. These are the times when it pays to have many paddocks so that the rest period is longer. A good practice is to have a sacrifice area in which animals can spend wet soil and slow-growth periods so that the grass is not damaged. This is the time to feed the hay you made when the grass was growing quickly.

Rotating livestock over many paddocks also has the happy side effect of helping break internal parasite life cycles. Intestinal parasites dropped in the grazing animal's dung are left high and dry when the animals have moved on and the sun beats down. Even if the weather is cool and the grass moist, leaving a higher stubble means that animals are less likely to ingest parasite eggs that may be on the ground or lower on the grass stem.

Minerals in the Soil

As your animals come to gain more and more of their food from your own pastures, it follows that their nutrition and thus their health—and your health if they contribute to your diet—need to be supported by the minerals present in your soil. Have a soil test done to see what you have to work with. Laboratory soil tests are all pretty objective, but the fertilizing recommendations that come with your results may reflect different goals, such as maximum bulk plant yield or maximum animal nutrition. I recommend educating yourself about how soil fertility affects the health of plants, animals, and people before deciding how to fertilize.

Rotational grazing works because it mimics the natural system of grass and grazing animals. Leaving an adequate stubble height and allowing grass to recover before being grazed or cut again will give your grass a competitive advantage over pasture weeds and increase the productivity of your pasture.

From *Ruminations* #54

The Ideal Pasture

The ideal goat pasture will include clovers and mixed grasses. It also should have plenty of twigs, branches, brush and trees. Goats are browsers and like a variety of items in their diet. They should not eat just grasses; in fact they love to choose among many foods.

If your pasture has no trees, bushes or other brush, you can supplement it with browsing materials. Cut grape vines and fruit and evergreen tree branches (avoid Ponderosa pine) and stack them for goats to snack on. (DO NOT feed branches of cherries or plums to goats because of the possibility of prussic acid poisoning.) Make sure not to use trimmings that have been sprayed.

Plant a hedgerow outside the fencing with plants that the goats can eat once the plants are large enough for them to reach. These can include bamboo, sunflower, and various herbs.

Get leftover pumpkins after Halloween and put them throughout the pasture for goats to eat at their leisure, seeds and all!

Supplements

For optimal health, goats need vitamins, minerals and other nutrients beyond those contained in their hay, grain and forage. Mineral mixes are the most common supplements, but I have also included a discussion of others that are commonly fed to goats.

Minerals

Minerals and vitamins are essential to goat health, affecting growth, fertility, skin health and kid development, among other issues. Some minerals are present in very small amounts in feeds, but most goats don't need a large amount of them. Goats with a variety of browse also may get more minerals and vitamins from their food than those that are housed in a small area eating just grass hay.

It's also helpful to know that goats will eat the minerals that they need—so make sure that they always have access to a good mineral block (as well as other supplements they need), along with their water and food.

How Do I Determine What Minerals to Use? Various kinds of mineral mixes are available for goat producers. Some contain not only trace minerals, but vitamins. They can be in blocks or loose. They can contain salt or be salt-free. They also will vary by the amounts of each mineral that they contain. Minerals with salt are more palatable to goats and save the goat keeper from having to additionally supply a salt block or loose salt.

Some goat keepers use a cattle or a horse mineral because goat minerals are not available locally. If this is necessary, make sure to check the feed tag (see p. 23) to determine the actual contents. (Goat minerals also may be obtained online, sometimes with no shipping charge.)

The choice of which mineral mix to use is up to the goat keeper, the goats' preference, geographic location and your specific setup. Regardless of the form chosen, minerals should always be available free choice to goats. (For goat keepers who have large herds, contacting the local extension office and working with a nutritionist to create a custom mineral mix may be more cost-effective.)

Many feed stores offer combination sheep/goat minerals, but we have learned over the years that these don't contain enough copper to meet the requirements of goats. In addition, if sheep are fed goat minerals, they may develop copper poisoning. Goat owners who house sheep and goats together will need to face the additional challenge of providing appropriate supplementation needed for each species, without harming the other.

Trace Minerals in Supplemental Mineral Mixes

Calcium (Ca) is essential for goat health. It must be supplied by feed. It is required for bone health and growth. Legume hays, such as alfalfa, are higher in Ca than grass hays such as bent grass or timothy. Pregnant does that get too much calcium may develop milk fever.

Phosphorous (P) must be fed in the correct proportion to Ca in the goat's feed. The ratio of Ca to P should never drop below 1.2:1. Like calcium, phosphorus is needed even more by lactating does and growing kids, at a ratio of 2:1 (2X Ca to P). Most grains are high in P and low in Ca.

Sodium (Na) and **Chlorine (Cl)** together make what we know as salt. Salt is important to many bodily functions.

Potassium (K) is present in fresh forages.

Iron (Fe) is important in blood and prevention of anemia. Small amounts are present in a goat's regular diet. Because milk lacks iron, kids that are fed only milk have been known to develop a deficiency.

Copper (Cu) is also necessary for maximal goat health. It is often found in soil, but in some areas it is deficient. Many breeders who have identified such a deficiency are now supplementing their goats with copper boluses. Copper works with molybdenum, another trace mineral.

Iodine (I) is used by the thyroid gland. It is deficient in some US soils.

Sulfur (S) is a component of many proteins. Rumen micro-organisms need it to build proteins.

Magnesium (Mg) is important for nerve and muscle function, among other things. Deficiency doesn't usually occur, except in fast-growing, pastures that were heavily fertilized with nitrogen (N) and potassium (K). This can lead to a condition called grass tetany. Too much Mg also can predispose a wether to urinary calculi.

Selenium (Se) Many soils throughout the US are lacking in selenium. (See map at www.saanendoah.com/map1.html.) Selenium works with vitamin E to prevent white muscle disease and retained placentas and to reduce susceptibility to parasites and other disease.

Zinc (Zn), **Manganese (Mn)** and **Cobalt (Co)** all are needed in trace amounts by goats. They normally can be found in the regular diet. Co is deficient in some soils. It is essential to synthesis of vitamin B12 in goats.

Mineral mixes are used primarily to supplement necessary calcium (Ca) and phosphorus (P), as well as to provide other macro and micro (trace) minerals, such as selenium, zinc, cobalt, iron and others. They also provide essential vitamins. Goat minerals should have a high concentration of vitamin A. Concentrations of vitamin D and E are then added proportionally to the amount of vitamin A. A good level of supplementation is up to 500,000 IU of vitamin A per kg of mineral.

To determine whether a mineral will meet your goats' requirement, read the feed tag. Minerals have a tag with two numbers on it, e.g., 18:18 or 18:9. These numbers identify the concentration in percentage of Ca and P in the mineral. The 18:18 mineral has 18% Ca and 18% P, while the 18:9 mineral has twice as much Ca as P—two-to-one. Probably the most ideal goat mineral is the two-to-one. The feed tag also shows percentages of trace minerals.

How Much Mineral Is Necessary? The amount of mineral a goat will eat will vary by goat. Since it is being fed free choice, you just need to make sure that it is available. Because good mineral blocks/loose minerals are so expensive, you may want to track how fast they are being eaten for purposes of budgeting.

Another consideration is keeping the minerals clean. Kids have a tendency to jump into feeders or on top of blocks and adult goats will poop or pee wherever they are standing, even if minerals are right behind them.

Baking Soda

Many goat owners offer free choice baking soda. It aids digestion and helps prevent bloat. Goats will use it if they need it. My experience is that it is often wasted, so just add a little at a time to the feeder.

Beet Pulp

Beet pulp adds fiber to the goats' diet. Some goat keepers add it to their feed ration, but I hydrate it in hot water and put it out in bowls for the goats, mostly in the winter when they spend less time outdoors eating other kinds of plants. At first they turn up their noses, but over time they grow to love it, fighting and slurping it down like pigs.

BOSS

Black oil sunflower seeds (BOSS) contain vitamin E, zinc, iron and selenium, and are a good source of fiber. They are commonly included with goats' feed as a nutritional supplement. Goat keepers claim that the fats make their goats' coats shinier, and they are said to increase the butterfat in their milk. I buy BOSS by the 50 lb bag and what the birds don't get, the goats have mixed into their grain.

Vitamins

Vitamin A is produced in the body from the beta-carotene in green plants. It is stored in the liver and the fat, so it can be used when green feed is lacking. Vitamin A deficiency occurs only when the goat is deprived of green plants, such as high-quality, leafy, fresh hay. Vitamin A is essential for eyesight, fertility and fighting infection. It also provides healthy skin and internal organs.

Vitamin B is manufactured by the micro-organisms in the rumen. This generally only becomes an issue if a goat develops a digestive problem such as acidosis, which can kill the healthy micro-organisms that make thiamine (vitamin B1). Thiamine deficiency is also called polioencephalomalacia.

Vitamin C (L-ascorbate) is an essential nutrient made by the goat's body. It also can be found in fruits and vegetables.

Vitamin D is produced by the goat's skin from sunlight. Goats that live in northern climates or are kept indoors need to have it supplemented in their diet. Fresh, sun-cured hay is a good source of vitamin D. Vitamin D is necessary for proper bone growth and health. Deficiency may be evidenced by rickets in goat kids or brittle bones in adults.

Vitamin E works with the selenium to ensure normal growth. A deficiency in the two of these can lead to white muscle disease in young kids.

Vitamin K is made by the micro-organisms in the rumen and often included in feed. It is essential to blood-clotting.

Kelp

Kelp is a seaweed in the algae family. This plant grows in the ocean, or a marine environment, and like other plants uses photosynthesis to produce food. It is a good source of nutrients, including minerals, iodine, selenium and other trace elements. In addition to being used as a plant fertilizer, it is also a good supplement for goats (and people, for that matter).

How to Feed. Kelp meal can be given to goats free choice, just like minerals or baking soda. Some people combine minerals and kelp into one source. Because it is high in selenium, it is especially good for pregnant does and young kids. The goats seem to know when they need it; I often can't keep up with their demand and they actually fight over it.

Pat Coleby, in her book *Natural Goat Care,* discusses the use of kelp meal to prevent and to treat various deficiencies. Because of its iodine content, she recommends that it be fed to any goat exhibiting the signs of iodine deficiency: in mild cases, dandruff, and in more severe cases, goiter (a swelling of the thyroid in the neck). She counsels against putting it into the feed though; she believes that goats eat what they need when it is served free choice.

Others suggest that with healthy goats, kelp should be mixed with other feed. This may be a better way to ensure that all of the goats get as much as needed. One company recommends 1 lb of kelp for every 50 lb of feed.

The addition of kelp to the diet may benefit dairy goats in many ways, including improved production (milk, butterfat, decrease in mastitis) as well as overall hardiness and health.

Where to Buy Kelp. Many feed stores carry kelp and others will special-order it if you ask. It can be expensive (depending on the number of goats you have) and the price varies with your location. Most of the kelp on the market is from Nova Scotia or Iceland.

Kelp meal can also be purchased online from various companies. One problem with online purchase is that with its weight, kelp meal can be costly to ship.

Avoid purchasing kelp supplements that are marketed for people, as the price will be exorbitant. Kelp is often marketed in weight loss remedies, as well as for use as a general supplement.

Kelp is a good investment for raising healthy goats. It supplies them with a lot of nutrition with little effort. And your goats will be happier and more productive in the long run, too.

From *Ruminations* #56

Apple Cider Vinegar

Some goat keepers add unfiltered apple cider vinegar (ACV) to their goats' water. Claims regarding its usefulness vary. According to the Bragg health products web site (www.bragg.com/products/vinegarPets.html), ACV contains "enzymes and important minerals, such as potassium, calcium, magnesium, sulfur, chlorine, phosphorus, iron, silicon and other trace minerals. The vitamins contained in ACV are bioflavonoids (vitamin P), beta-carotene (precursor to vitamin A), and vitamins C, E, B1, B2 and B6. Tannins from the crushed cell walls of fresh apples as well as malic acid, tartaric acid, propionic acid, acetic acid and pectin (fiber) are also contained in ACV."

Assorted Treats

Goats don't need a lot of treats, but some of us can't help but ply them with snacks. Despite the claim that goats will eat anything, most of them are actually quite picky.

Some snacks that they like and that are good for them include carrots, lettuce, bananas (peel and all, if organic), apples, pears, melon, oranges and plain, salted corn chips.

Tip: When feeding apples to goats, make sure to cut them into small pieces. Otherwise, they can get stuck in the throat of an overeager goat, causing it to choke.

Avoid members of the nightshade family (such as potatoes) and kale or other plants with oxalates. If you have a question about a plant's toxicity, check it online at www.ansci.cornell.edu/plants/.

For a kelpy goat treat, try this recipe from Maine Sea Coast Vegetables, in Franklin, Maine.

Candied Kelp

12 inches dried kelp
¼ cup honey
½ cup water
1 cup sesame seeds

1. Soak kelp in water until very soft. Cut into desired shapes, enough to fill ½ cup.

2. Bring honey and water to a boil.

3. Reduce heat, add kelp, and simmer uncovered until almost all liquid has been absorbed or evaporated (1 to 1½ hour). Check frequently, adding a dash of water when needed and stirring occasionally.

4. Arrange kelp pieces on sesame seeds, turning to coat.

5. Bake on a clean cookie sheet at 300° F., for 25 to 30 minutes. Halfway through baking, turn over the pieces, being careful not to scorch the sesame seeds.

Note: Maple syrup may be mixed with the honey for a different taste. You may also replace sesame seeds with ground almonds, pecans, walnuts or peanuts.

Hoof Trimming

Hoof trimming is an essential part of being a good goat keeper and maintaining goat health. It is easy to avoid or put off, but such procrastination can have a negative impact for a long time. Trimming a little bit frequently is better than trying to trim a lot at one time. Frequent trimming also helps to keep the hooves in the right shape, rather than having to correct misshapen hooves later on.

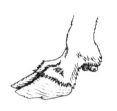

Why trim? Goats with misshapen, overgrown hooves may not be able to walk properly and, as a result prefer to lie down. This can lead to missing feedings and an inability to graze effectively, breed or even properly mother their babies.

Hoof care is particularly important during wet weather or in wet climates, due to the risk of foot rot. This is a bacterial infection that can endanger a whole herd, as it spreads in wet, muddy areas. Untrimmed hooves are more susceptible to this condition. Treatment is a lot more time-consuming than routine hoof trimming.

What do I use to trim hooves? Many different types of hoof trimmers are available. Check goat catalogs and feed stores, and ask other goat owners what kind of trimmer they prefer. Possibilities include tin snips, pruning shears, hoof trimmers, and heavy scissors. My experience has been that paying a little more is worth it in the long run; the cheap hoof trimmers wear out sooner. In addition, they will need to be sharpened from time to time, and some of the trimmers have a limited lifespan in terms of sharpening.

How often should I trim? A variety of factors will determine how frequently you should trim hooves, including age (yours and the goat's), environment, housing, breed, level of activity, nutrition and conformation. Check your goats' hooves every few weeks to determine the best schedule and record the trimming on the goat's health record. Pretty soon, you will get into a schedule. And remember, you generally can't trim too frequently. If I am feeding goats on a stanchion, I routinely check the hooves.

How do I trim the hooves? When trimming a kid's hooves, you can simply hold it on your lap and fold the front leg back. The back leg is more challenging; you can hold it out, being careful not bend it, then trim the hoof. I flip the goat around facing the opposite direction after trimming the front and back hooves of each side.

For adult or large goats, restrain the goat by putting it on a milking stand or positioning it next to a wall with someone else's help. Gently bend the goat's leg at the knee to minimize struggling. Starting with a front foot gives the goat a chance to see what is being done, possibly making the chore easier.

Carefully use the point of the hoof trimmers to clean out dirt or manure that has accumulated between the hooves or caked on the sole.

After cleaning the sole, trim off the wall of the hoof starting at the toe, so that it's even with the sole. Trim away any hoof wall that is folded over so that it is level with the sole.

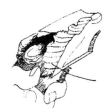

Be sure to trim around the entire toe of the hoof. If rotten tissue has developed between the soles of the hoof, trim that as well, being careful not to cut too deep.

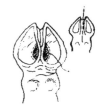

Trim the heel so that it is even with the sole. Watch closely—as you get closer to living tissue you will notice that it is pink colored. In cases where the hooves have not been trimmed for a while, a pocket of dirt or foul-smelling tissue may be present. If this is the case, clean the area by cutting it away and then clean the area with a disinfectant such as Betadine® soap.

When you are finished, your goat will have even, flat hooves that allow it to walk properly and distribute its weight appropriately.

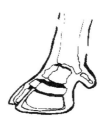

When first learning to trim a goat's hooves, trim only small amounts at a time. This will allow you to be more detailed about getting your goat's feet properly trimmed and will allow you to trim your goat's hoof without getting too deep into the sole, causing it to bleed.

As your confidence develops, you will invariably trim a hoof too deep, causing the goat to bleed. You can put blood stop powder on it, but usually just the pressure of standing on the foot will stop the bleeding. Another option is to keep a disbudding iron plugged in during hoof trimming, so it can be used to cauterize a wound to stop the bleeding.

Trick: In a pinch, apply a wad of spiderweb to a cut hoof. It contains a substance that is effective in stopping bleeding and preventing infection.

Make sure to disinfect your hoof trimmers after each use.

Can I do anything to naturally help the hooves? Always house goats in a dry area to prevent foot rot or foot scald. Adding a large play rock to the goat's pen is one way to help them naturally wear down their hooves.

FAMACHA

Anyone raising goats, particularly in warmer climates, needs to be trained in FAMACHA in order to keep a deadly parasite under control and keep their herd healthy.

FAMACHA is a system for controlling *Haemonchus contortus*, a gastrointestinal parasite also known as the barber pole worm because of its red and white stripes. The main health consequence of this parasite is anemia. Kids and pregnant or lactating does are at the highest risk.

This system provides a tool for goat keepers to use in assessing the level of anemia in individual goats and to selectively deworm those that are affected. The intent is to keep goats healthy while trying to limit the growing resistance to dewormers that has come about from overuse of these products.

Parasitologists now believe that the most significant factor in the development of resistance to dewormers (anthelmintics) is the practice of deworming all goats in a herd at the same time. This is because only the resistant parasites survive, going on to reinfect goats through eggs in feces left in the pasture. In addition, only about 20–30% of animals have 70–80% of the parasites, so under this old method many goats that are receiving treatment don't really need it. A third factor that has led to dewormer resistance is the practice of "rotating" wormers, so that a different one is used at a certain interval—for instance, every three months.

> **Tip:** In some cases, anemia may be caused by a condition other than *Haemonchus contortus*, such as:
> - Liver flukes (more likely in gulf coast and NW states)
> - External parasites, such as sucking lice
> - Parasites of the blood (uncommon in the US)
> - Bacterial or viral infection
> - Nutritional deficiency

The FAMACHA system was developed by the South African Veterinary Association and was modified (and now promoted) by the Southern Consortium for Small Ruminant Parasite Control (SCSRPC) (www.scsrpc.org), under a federal grant.

FAMACHA consists of looking at the color of the lower eyelid (a mucous membrane), and comparing it to colors, assigned numbers 1–5, on a small laminated card. Colors range from red (healthy) to almost white (anemic). The lighter the color, the more anemic the goat and more likely it has a problematic level of the barber pole worm and needs to be dewormed.

Does become more susceptible from two weeks before their kidding date, so should be closely monitored during this time.

Animals also need to be checked for "bottle jaw," a soft swelling under the jaw that is indicative of this parasite. Even if the goat's anemia level under

Goat Health Care

FAMACHA doesn't indicate the need for deworming, it should be wormed anyway, as bottle jaw is another serious symptom.

The system has a number of advantages, in addition to decreasing parasite resistance. It can lead to decreased costs for goat keepers, who are no longer deworming all of the animals. It also will enable the goat keeper to determine whether a dewormer is not working, if no change is seen after treatment. If animals are kept in different areas, the system can also help to identify which areas have more parasite buildup so that the animals can be rotated. It also has the side effect of identifying other problems that the goats may have, when the routine exam is done. Goats that consistently show problems can also be identified for culling.

Only veterinarians or goat producers who have attended a formal training workshop presented by a veterinarian may obtain a FAMACHA card. It is essential that everyone who is implementing this system use the card for diagnosing their goats according to instructions in the guide that comes with it.

> **Tip:** SCSRPC recommends what they call "Smart Drenching." This includes:
> - Administering dewormer at *two times* the cattle dose—except levamisole and moxidectin—for the heaviest animal,
> - Using injectable and pour-on products *orally*
> - Delivering the drug over tongue into the pharynx
> - If not on the FAMACHA program, using one class of dewormer for at least a full year or until it doesn't work **(do not rotate monthly)**
> - Putting animals on a dry lot for 12–24 hours and not feeding them (fasting) before deworming, and repeating administration in 12 hours.

It is important to note that this system is effective only in controlling the barber pole worm and *not* other gastrointestinal parasites, so a program should be in place for monitoring through fecal testing and treating them as well, using the same principles of limiting dewormer usage to what is necessary. Different dewormers may be required to get at these parasites. For example, Safeguard and Valbazen are effective against most lungworms.

To learn more about FAMACHA, see the SCSRPC web site at www.scsrpc.org, or have your veterinarian or extension agent send an e-mail request for more information to: famacha@vet.uga.edu. Veterinarians can purchase FAMACHA cards by making requests using this same address.

If training is available in your area, make sure to attend. It is well worth the time and money, and the system goes a long way in keeping goats healthy, if used properly.

Creating "Super Worms"

In a Guest Editorial in July/August 1999 *The Goat Farmer* magazine, Ken Sykes gives a recipe for breeding superworms. He points out the major problems for what he calls the "commercial" and the "alternative" enthusiasts. The commercial goat farmer gives little attention to long term effects, side effects and reduction in efficiency over time of using drugs that have been researched. The alternative farmer has a problem with "sifting the chaff from the wheat." That is, since there is only anecdotal testing of these methods, it is hard to know what really works.

Tip: Do not try to eradicate ALL worms in your goat. Deworm only for an increased number of worms. If the goat is not thin, has no diarrhea and looks happy—don't deworm unless a fecal exam shows parasites.

He goes on to list the ingredients useful to breed a superworm: reasonable rainfall and a temperature range suitable for worms; high pasture that allows the worms easy access to their hosts; high numbers of goats; and goats that are particularly susceptible, such as pregnant does and kids. Next, dose the goats with a chemical at a low rate, then a higher rate after the weaker worms have died; change chemicals often, and introduce many new goats from different herds. Finally, fail to provide additional vitamins and minerals.

When the problem seems overwhelming, get out of the goat business dispersing your herd as widely as possible, so as to introduce the parasites into as many other herds as possible.

Sykes' point was that total herd management is the real answer to controlling worms. Some may choose chemical means and some may choose holistic. The ultimate goal should be not only to make goat farming sustainable, but to improve it. Probably the most important factor in this equation is to see the whole picture and learn from it. Find out what works and do it.

Parasite Control

by Sue Reith

For a long time now I've been encouraging goat owners to take charge of their own animals' health management. To this end, simply because it's such a wonderful diagnostic tool, I provide instruction below in how to check animals' fecal samples to determine what, if any, parasites they're harboring. By checking out the fecals you can tell very quickly if there are worm eggs present, and if so, exactly what type they are. This is important because while most worms can be controlled with a good general wormer, there are some very damaging species that will only respond to one specific type of medication. Also, sometimes in doing a fecal check (particularly in very young or debilitated goats) the sample may reveal that the culprit causing the animal's difficulty is coccidia, instead of, or in addition to, worms, in which case a treatment other than worming will be needed. And last but not least, if a sample check shows there are neither worm eggs nor coccidial oocysts (coccidia eggs) present, you won't need to treat unnecessarily. In addition to that being better for your animal, it is also very cost effective!

Here's the recipe for checking the status of your own livestock fecals. (You can also do them for your dogs and cats!)

First, you need access (preferably in your own kitchen) to a decent microscope. It doesn't need to be brand new, or high-tech, but should have its own light source, and a "moveable stage," as well as 10X and 40X power. You will need glass slides and covers to go with it. Often used microscopes of good quality can be found in a college newspaper ad, and even in pawn shops.

Second, it's absolutely essential to find a good (and complete!) set of clear photos to identify the various eggs and oocysts you find. The only book that I know to be a good reference for use while actually examining the slide is the *Veterinary Clinical Parasitology,* any one of the 1st through 5th editions, by either Benbrook and Sloss or Kemp and Sloss. This book is an excellent quick on-the-job reference, but sadly those 1st through 5th editions are now out of print. There is a 6th edition of the same book currently on the shelves, but it has unfortunately been "revised" by one Anne Zajac, who essentially turned it into a sort-of primer for use apparently in a beginning parasitology class, completely ending its usefulness as an aid for identifying the eggs/oocysts discovered on the slide.

In my quest for a substitute for the *Veterinary Clinical Parasitology* book I've purchased both Foreyt's *Veterinary Parasitology* and Georgi's *Parasitology for Veterinarians,* and have checked out vet school bookstores on various campuses to see what they have available. While most are very informative, the 1st through 5th editions of *Veterinary Clinical Parasitology* continue to provide the only complete, on-the-job reference pictures available for actual use at the side of the microscope as you work to compare pictures with parasite eggs/oocysts to identify what you see on the slide. In that book, clear, sharp pictures of the eggs and oocysts found in feces are categorized by species of animal host, so that there is a whole section on eggs/oocysts found just in dogs feces, and another section on chickens, and another on goats, and so on. On each page there is a photo of several eggs together, as you might see on your microscope slide, and then a close-up photo of a single egg of the same species, to help with details.

> ## Multi-Vitamin & Copper for Parasite Control
>
> The results of two experiments reported in the October 10, 2006, issue of *Veterinary Parasitology* suggest that the use of a sustained-release multi-trace element/vitamin (TEB) containing copper may be effective in controlling gastrointestinal parasites in mature goats. The main parasite affecting the goats was *Haemonchus contortus* (barber pole worms).
>
> In both of the experiments, which were carried out in the summer and in late gestation, goats treated with TEB showed a lower worm burden in the goats that were treated versus those that were not. The effect did not last more than 28 days, however, so additional control measures may be needed.
>
> From "Health and Science News,"
> *Ruminations* #55

Other parasitology books provide pictures of the larval and the adult forms of the parasite, but when doing a fecal exam the only thing you see on that slide is the egg or oocyst. The larval and adult stages are busy doing damage in other parts of the body (e.g., lungworms in lungs, heartworms in heart, strongyles in intestines), or perhaps still in the vector outside the body, thus not in the feces. The fact is that you simply can't identify the kind of egg or oocyst you are looking at on the slide unless you actually have a good, clear picture to compare it with. And it goes without saying that if you can't accurately identify the egg/oocyst, then you won't know what treatment is best for wiping it out.

I urge you to try to locate, in a used educational bookstore or on the Internet, a copy of one of those early editions of *Veterinary Clinical Parasitology*. Several excellent book-oriented search engines provide access to used and out-of-print bookstores' inventories…but the current high demand for those 1st through 5th editions of *Veterinary Clinical Parasitology* by so many people who want to take charge of this aspect of their animals' health management makes dedication to the search essential, because when they become available on the Internet they are snapped up in a hurry. EBay (www.ebay.com) occasionally has one up for auction, but due to the upsurge of interest in learning how to check livestock and domestic pet fecals at home, the price for it has skyrocketed.

Because it's so hard to find a copy of those 1st through 5th editions of *Veterinary Clinical Parasitology,* I've scanned the goat oocysts/eggs to help people get started while looking for one. Contact me at suereith@msn.com and I'll be happy to e-mail the picture attachments to you for use while you continue your search. Additionally, the A Pacapacas Farm web site now has made the wonderful Paul Miller (Ocicat) page, "ParaSite," containing pictures sorted by host species of eggs and oocysts of goats, sheep, cattle, cats and dogs, available at www.apacapacas.com/parasites/, for which we are eternally grateful!

Third, either buy, or make up, a quantity of egg flotation solution, and locate some Fecalyzer containers in which to prepare the samples. Get some microscope glass slides and covers as well; you will need a small box of each. (The Revival catalog, www.revivalanimal.com, or 1-800-786-4751, has the Fecasol, fecal flotation system tubes, glass microscope slides, and glass cover slips available both in their catalog and on their web site.)

In using a special fecal solution the basic idea is that the solution must be heavier than the eggs so the eggs will float to the top, as this is a "flotation" method of producing a slide sample to view under the microscope. Flotation solution can also be homemade using the following methods:

To make a saturated solution of salt (table salt) for fecal flotation take a glass container with a good closure and fill it partially full of water (tap water will do nicely). Add a quantity of salt to the water. Add enough salt so that undissolved salt remains in the bottom of the container. Check and stir the solution in the container every now and then and add more salt if there isn't a generous layer of salt still on the bottom. You can continue the process at room temperature over a period of 24 hours or so, or speed up the process by heating gently. The liquid is considered saturated when no more salt crystals will dissolve in it with continued stirring. The fluid is then withdrawn from the mixing container, leaving the undissolved sediment behind. The saturated salt solution is now ready to use in preparation of the sample for placement on the slide. Excess solution can be stored in the refrigerator for further use in the near future. One caveat regarding the salt solution: It has a tendency to dehydrate the fecal sample specimen if you don't work quickly.

Saturated sugar solution can be made the same way as salt solution. A suggestion has been made that a small amount of formaldehyde added to the sugar solution will keep it from spoiling. This saturated solution should also be refrigerated for further use in the near future. You don't need any special equipment to make these saturated solutions since the undissolved material at the bottom of the container is your proof that the solution is saturated. I am told that a saturated solution can also be made of magnesium sulfate (Epsom salts) in the manner previously described and that it, too, works well for flotation.

Fourth, gather all the materials together, and collect a small amount (about ¾ teaspoon?) of fresh, normal-consistency poop (or fecal matter, if that sounds better to you), place it in the fecalyzer container, pulverize (mash) it, and dilute it (I stir it with a swizzle stick) with flotation solution. Fill the container to all but overflowing, blot off any foam that rises to the surface, and then carefully cover it with a tiny glass cover slip. Wait about 20 minutes to give the eggs plenty of time to rise to the top of the solution, and then carefully lift the cover plate off, keeping it level so that a drop of liquid remains on its underside. Set it on the glass slide, place the prepared slide under the lens (which is set on the lowest microscope power—10 X 10), open the book to the chapter covering parasite eggs/oocysts found in goats/sheep (great visuals there), sit down, and enjoy the view! It helps if you enjoy the challenge of finding things under a microscope, or have a good friend who does and will guide you.

> **Tip:** If you aren't able to do your own fecal exams, you can instead collect the feces and send them to a veterinary lab immediately or take them to a local vet who does such testing. If they can't be sent immediately, refrigerate (do not freeze).
>
> Submit the samples in a container (box) in individual, sealed, double Ziplock bags labeled with your name or farm name, the animal species, and the animal number or name. To ensure freshness, include blue ice in the container and ship for delivery within 24-48 hours.

Fifth, once you have discovered which parasite(s) are present, do your research into the various wormers/coccidiostats, etc., and find which one(s) are most effective in treating the specific parasites you've found in your animals.

Sixth, find and administer the correct product!

Note: The materials needed, including the book but not including the microscope (which depends on how good a deal you can get elsewhere) should cost in the neighborhood of $50. That's roughly what four to five fecals done professionally would cost. (Editor's Note: In 2008 a fecal exam for parasites at the Washington Animal Disease Diagnostic Lab [WADDL] cost $18, with an additional $10 accession fee). The best part is that doing them yourself gives you a wonderful sense of accomplishment and independence!

From *Ruminations* #56

Giving Injections

The new goat keeper can put off learning several unpleasant procedures, such as disbudding and castrating, for a long time. Since most people vaccinate their goats at least annually, and antibiotics must be administered by injection, most goat keepers find that they must learn to give shots within a year of getting their first animals. A person who is squeamish will find this a daunting task; with a bit of coaching and practice, giving shots need not be traumatic for the goat or the owner.

After determining what medication to give a goat (which may involve consultation with a veterinarian and/or a knowledgeable breeder), the next item of business is to decide what type of injection is right for the situation. The first concern may be rapid absorption, or choosing a type of injection that can be repeated frequently without causing too much discomfort. Choosing the right type of injection is key to its effectiveness.

Types of Injections

The medication vial usually will include directions for administration. However, most medications that can be given intramuscularly (IM) also can be given subcutaneously (SQ). Most medications are given SQ to goats, particularly those consisting of large quantities and those that must be given frequently.

SQ injections go between the skin and underlying tissue. This injection method provides the fourth fastest rate of absorption. The site for an SQ injection is generally the neck, behind the shoulder, or anywhere over the rib cage, not too close to bony structure such as the backbone or shoulder blade.

Injection Sites

IM medications, given in the muscle, provide the third fastest absorption rate. Many sites are appropriate for IM injections, but the easiest and safest are on the top of rump, in the hip, or on either side of the backbone in the neck region.

SQ and IM injections are the most common methods, and are the ones used for vaccination and administration of antibiotics or a pain medication such as Banamine.

Other types of injections are intraperitoneal (IP) and intravenous (IV). IP injections are made directly into the peritoneal cavity, on the right side of the goat, behind the rib cage and in front of the hip bone. This method provides the second fastest rate of absorption into the system. It is also used for administering large volumes of medication, but generally is not recommended due to the danger of piercing internal organs, with a risk of peritonitis resulting from the medication used.

IV injections go directly into the vein. The jugular vein in the neck is the most suitable location for an IV injection in dairy goats. The rate of absorption of medication given by this method is the fastest. It must be used for administering medications that irritate tissue when injected by other methods.

How Much to Give

Once the injection method is chosen, the dosage must be determined. At times the dosage for dairy goats differs from that recommended on the bottle, both in volume and in length of treatment. This is because most of the products developed and labeled for livestock have not been tested on goats and are not FDA-approved for their use. Using such a product in a goat is considered to be "extra-label use." The fact that a product is not labeled for goats does not mean that it is unsafe or ineffective in them; just that testing on the species has not been done. Much of the problem is due to the relatively small number of goats in the US. Drug companies cannot economically justify going through the testing required to approve a drug for use in goats (considered a minor species). As a result, experience, coupled with information from veterinarians and breeders, determines the effective dose in caprines.

For some drugs and vaccines "one size fits all"; for others, dosing by weight is essential. Please read the suggested dose material to determine the correct dose for your goats. Always check the bottle labels for strengths to avoid over- or under-dosing (many drugs come in more than one strength), as well as for additional dosing information, precautions, indications, withdrawal times, etc. Milk withdrawal and discard times are for FDA-approved dose rates and, unless otherwise noted, for cattle. Once the dosage is determined, writing it on the bottle for future reference is helpful.

Giving the Injection

Disposable syringes and needles are available at feed stores, mail order veterinary suppliers and veterinarians. Syringes can be found in one, three, six and 12 ml sizes. The 3 ml and 6 ml are the most practical to have on hand. Needles are sized by length and thickness. One inch needles in 16, 18 and 20 gauge are most commonly used for goats. The higher the gauge, the thinner the needle. After using a needle, cap it with the plastic sheath and throw it away. Always use a new needle for each goat to avoid disease transmission.

Once everything is ready, it's time to administer the injection. Having a vet or an experienced goat keeper demonstrate how to give a few shots and then observe your first shot is ideal. For those who have not previously given injections, a needle and syringe can be sacrificed for a practice run. Fill the syringe with water and practice sticking the needle into an orange (or piece of bacon with the skin on), injecting the liquid by depressing the plunger.

Before giving the shot, wipe the top of the medication bottle with alcohol. Insert the needle into the bottle and withdraw the necessary medication, tapping the syringe occasionally to get the air bubbles out. Once the medication is drawn, tap the syringe again and push out any bubbles.

Immobilize or have someone hold the goat and locate the area to give the injection. Wiping the area with alcohol is not necessary. For an SQ injection, lift the skin into a tent. Stick the needle into the tent, making sure it is not in the skin, the muscle or through the other side, then inject the medication and withdraw the needle.

For an IM injection, select the site for injection, stick the needle into the muscle, withdraw the plunger slightly to ensure that you have not hit a vein or vessel (if you have, blood will be drawn into the syringe and you should adjust the needle or pull it out and start over again) and depress the plunger slowly. Withdraw the needle and rub the area firmly but gently. For either type of injection, the more it is done, the easier it gets.

Medications may lose potency over time. While they may not become ineffective after the expiration date (this varies by medication), paying attention to expiration date will let you know that they need to be replaced. A medication that has become discolored or cloudy or contains a sediment should be discarded and replaced with fresh medication.

Most goat keeping tasks seem difficult at first and giving shots is no exception. With practice and patience, giving routine injections to goats will soon seem like second nature.

From *Ruminations* #57

Helpful Hints Regarding Injections

Always have epinephrine on hand when giving injections, in case of an anaphylactic reaction.

Anaphylactic shock is a severe, immediate allergic reaction. If a goat suddenly collapses or goes into shock after a shot, administer Epinephrine immediately IM or SQ at 0.1 ml (kids) to 1 ml (mature bucks).

The usual dose for ruminants is 0.5–1.0 ml/100 lb body weight SQ or IM of the 1:1000 strength. This may be repeated at 15 minute intervals as needed.

Drawing Blood

Goat owners with a large herd may find that having a veterinarian make a farm call and draw blood samples is too expensive. Those who live near a veterinary school may be able to have it done by students at a lower cost. Another option is to have a friend who is a vet tech draw the blood. Those who don't have any of these options available should consider learning to draw their own blood samples for routine testing.

The most common disease that goat keepers test for is Caprine Arthritis Encephalitis Virus (CAEV). Veterinarians recommend that whole herd testing be done every six months for the first year and annually after that. Several other diseases require blood for testing, including caseous lymphadenitis (CLA), brucellosis and Johne's disease. For CAEV testing, you will need 3 ml of blood. (Check with the lab where the sample will be submitted for required Vacutainer tubes and amount of blood to draw for other tests.)

Charges for such testing will vary by lab and test being performed. The Washington Animal Disease Diagnostic Laboratory (WADDL) in Pullman, Washington, charges an accession fee for a group of samples and then another fee per sample, with a discount for residents of the state. In addition, there are shipping costs. (Complete information for WADDL is available at www.vetmed.wsu.edu/depts_waddl/)

Items needed

Gather the necessary items, ensuring that you have enough for all goats to be tested. Include a few extra needles, in case you need a new one due to problems hitting a vein or keeping the needle in the vein to withdraw the full amount of blood.

- Rubbing alcohol and cotton balls, or alcohol prep wipes
- 3 ml syringes
- 3/4" x 20 mm needles
- Vacutainer tubes ("red-top" or serum separator [SST] tube)
- Sharpie marker for labeling tubes
- Clippers for shaving neck area
- Animal identification form
- Pen for recording on animal identification form

Vacutainer tubes may be obtained from a veterinarian, a friend, a medical or veterinary supply store (see Resources) or sometimes even on eBay or Craigslist.

Label each tube with the goat's name or number, the date and your name or farm name. Make a list identifying each goat by number assigned, if you choose to keep the goat names confidential.

The Procedure

Catch the goat whose blood is to be drawn. Have one or two people hold the goat—one of them holding the head area. One person can hold the nose with one hand and around the chest with the other. The other person (if necessary) can get behind the goat to keep her from backing away.

Find the jugular vein on the left side of the goat's throat. This can be done by placing a finger near the bottom of the neck and pressing—the vein will pop up when pressed. Press lightly with the other hand and you will feel it spring back. If necessary, shave the neck area to more easily locate the vein.

Remove the needle cap and insert the needle upward into the skin and vein at an angle nearly parallel to the vein, being careful not to push the needle all the way through the vein. Test to determine whether the needle is in the vein by gently pulling back on the plunger—if blood is in the syringe, you are in the vein. If so, continue to pull the plunger until the syringe measures 3 cc. If not, try again. (If you have to try too many times, replace with a clean needle so as not to introduce bacteria.) Replace the needle cap, put pressure on the goat's neck for about 30 seconds and then let the goat go.

Take the syringe of blood, remove the cap and insert the needle into the Vacutainer tube. Record with the goat's name or number and note it on the identification form for that tube.

Handling Samples

When finished with all animals, prepare the samples for shipment to the laboratory. If the samples cannot be shipped immediately, refrigerate them. Check with the shipper (USPS or FedEx) or the lab you are using for shipping instructions.

Vaccinations

Most veterinarians and goat keepers recommend vaccinating goats with *Clostridium Perfringens* type C & D-Tetanus (CDT) vaccine. Other vaccines are available for goats, both approved and extralabel. These include rabies, caseous lymphadenitis (CLA), pneumonia and others.

Goat keepers should talk to veterinarians and other goat owners to determine whether to vaccinate and what vaccines to give.

> **CDT Vaccination Schedule**
> - 1 month old, booster at 2 months
> - 1 week, then 1 month in problem herds
> - Use C/D +/- tetanus (not 7- or 8-ways)
> - 2–3 vaccinations/year

Vaccine Notes

Each vaccine has specific storage conditions. The majority of vaccines are best stored in the refrigerator where they are kept cold but not subject to freezing and thawing. (Freezing may reduce the potency of some vaccines and may cause local reactions at the injection site.)

- Killed vaccines, if properly stored and handled, may be used for the rest of the season or until expiration date. Note that vaccine labels recommend that the vaccine either be completely used or discarded after the vial is opened.

- Always use a new sterile needle to draw from the vial. Keep the vial refrigerated or packed in cold bags all the time. DO NOT allow it to become warm. Live and modified live vaccines that need to be reconstituted must be used all at once.

- Do not use chemicals or disinfectants when sterilizing syringes, needles or skin for administering live vaccines.

- Do not mix vaccines. Production of combined vaccines requires a very careful balancing of the components. However, more than one vaccine may be given at the same time, using separate syringes. These should be administered to separate parts of the body, at least 15 cm apart and preferably on different sides of the animal.

- Do not delay booster shots. Don't delay booster vaccination shots too long after the first shot, or you will lose the benefit of the first shot. The first injection is designed to provide a primary immune response. The booster shot gives a secondary response.

The primary response and secondary response work in tandem to provide maximum protection. Most of the immunity results after the second injection. The primary response of the first dose is fairly short-lived. When you wait too long to give the booster shot, the primary response may already have dissipated. In that case, the booster will actually provide another primary response instead of a secondary response.

Cold Weather Goat Care

by Cheryle Moore-Smith

We raise Nigerian Dwarf Dairy Goats in southern Maine and would like to share with you our list of things that we believe are necessary for us to do as good goat keepers in order to make our animals as comfy as possible through a long New England winter.

1. **A good supply of HOT water.** Our goats prefer to drink water so hot that you would think their little tongues would be scorched, but it is important to remember that a goat's normal body temperature is hotter than a human's. Even goats that have water bucket warmers will appreciate their morning and evening dose of hot water. This is especially important to those of you who keep bucks or wethers. Bucks and wethers can be prone to urinary calculi. Keeping your bucks and wethers supplied with hot water will ensure that they drink a lot of water; therefore they will have less of a chance of building up crystals in their urine.

2. **A constant supply of hay.** Hay should always be considered the goat's "staple" and any grain should be considered a supplement to their hay. You can expect to be filling hay feeders more often during colder weather so that your goats can keep their rumens active and healthy, which will help them to produce body heat.

3. **Extra grain.** I like to keep my goats looking and acting healthy on as little grain as possible, but when we get into the "real" winter weather (usually the time between Thanksgiving and Easter) I find that I need to increase the amounts of my goats' grain ration by as much as three times what I would be feeding otherwise. It is important for buck and wether own-

ers to remember that it is crucial to urinary tract health to feed either a grain that contains ammonium chloride or buy ammonium chloride separately and sprinkle it over the grain or mix it in the water.

4. **Shelter that is free from wind.** This is a very important part of keeping your goat healthy in the winter. They really are hardy creatures and can

stand amazingly cold temperatures as long as they do not have to survive the winds too. In making a shelter free from wind don't go overboard by making the shelter too insulated or airtight, as it is just as important that the shelter be ventilated. If a building is too airtight it soon becomes an unhealthy situation due to ammonia from urine.

5. **A dry place to bed down.** Be sure to keep your goats' sleeping quarters supplied with clean, fresh, dry bedding. Wood shavings or straw work well to absorb urine. If your building has wooden floors you should try to clean the bedding out every week rather than letting it build up (which would only be appropriate in a dirt-floored building). If your building has dirt or concrete floors, provide your goats with sleeping shelves, which can be something as simple as an old wooden wire cable spool. We also get free wooden shipping pallets and fill in the space between the boards with more boards. These act as little wooden stages that the goats can use to get off of the damp floors. They are easy to move around and clean, too. If you are wondering about when it is the right time to clean out soiled bedding just scooch down (in order to be at the same level as your goat's nose) and take a deep breath. If it smells bad or is painful to your lungs then you know that it will be just as offensive to goats, too.

Trick: Get down on the goats' level and take a deep breath to determine whether the barn needs to be mucked or have new bedding added.

6. **Regular parasite control.** This is very important! A goat that is carrying a large load of internal parasites is easily overcome by the cold because it will be immune-suppressed and most likely underweight from poor nutrient absorption. When in doubt, have a vet test a fecal sample (an easy and inexpensive thing to do to ensure the health of your goats).

Don't forget that your goats may easily be carrying external parasites too (mites are not species-specific so goats may get them from another animal that passes through their living space). Some external pour-on wormers work fairly well at controlling external parasites. Just remember that a whole herd can be carrying parasites, internal or external, and maybe only a handful showing symptoms. You still must treat the whole herd and not just the animals showing symptoms or you are wasting time and money.

7. **Good hoof care.** A good hoof care routine is always important, but even more so in cold weather. If the goat's hooves are not kept properly trimmed, debris can collect and freeze not only in each hoof clove but also between the cloves of each hoof. I tell new goat owners that they can get away with hoof trimming approximately every six weeks, but if they want to feel really good about the hoof care they are giving their goats they will want to trim monthly. During the winter months you should definitely trim every month.

Goat Health Care

I write down everything I do maintenance-wise in a notebook that I keep specifically for my goats but if you just want to go pick up a hoof to see if it is time for a hoof trim, always remember to check the back hooves and not the front. Goats tend to paw at things and will keep their front hooves more worn down than their back hooves.

8. **A place in the sun.** All goats are sun worshippers so if they have a spot that is up against the wall of a building that has southern exposure where they can retire to ruminate after filling their bellies with hay they will be very happy on sunny days.

9. **An exercise area.** With all the snow that we had earlier this season, it was very hard to keep up with the snow removal of just our long driveway! When you add up all the paths to goat pens that needed shoveling and gates that needed to be kept clear of snow, we had a full time job moving snow. It would be so easy to "cut corners" in order to save our backs but it is so important to remember to clear a large enough area in your goats' pen so that they can move around and get some exercise. If you don't clear an area free of snow in their pens, your goats won't voluntarily wander out in the snow, and therefore they won't be getting an acceptable amount of exercise. This is especially important to those pregnant does.

10. **Free choice "goat-specific" loose minerals and sodium bicarbonate.** This is another of those things that you should be doing year round for your goats but is even more important in the cold weather months.

11. **Set aside time to goat gaze.** This is the last on my list, but it may very well be the most important part of being a good goat caretaker. This is the part of goat owning that I enjoy most. It is when I remember just why I love having goats in my life. On even the coldest or stormiest day you will find me out with my goats (cup of coffee in hand) just watching and admiring their beauty and oftentimes getting some good laughs too. Doesn't sound hard does it? BUT it is so important to my herd's health. By watching my goats each day, I learn what the normal behavior of each personality in my herd is. This way when someone is even slightly "off" I am able to pick up on it before it becomes a big problem. So, don't forget to goat gaze!

From *Ruminations* #41

Breeding, Pregnancy and Kidding

Kids are one of the most fun aspects of goat ownership. There is nothing cuter than a bunch of kids joyfully bouncing around the pasture (or furniture)! While the vast majority of births occur with no problems, some knowledge and preparation are in order to ensure the best outcome in every case.

For instance, did you know that goat kids as young as two months old can reproduce? That's one reason that bucks are usually kept separate from does. It's also why bucklings need to be watched closely and made into wethers or separated at an early age. Doelings that are only a few months old still have a lot of growing to do and don't need the stress of kidding at seven months of age.

So what is the right age to breed a doe? The rule of thumb—for standard, not miniature goats—is when they reach 80 pounds, or around seven months of age.

Most goats are seasonal breeders and come into heat in the fall. A doe will give several different signs that she is in heat. She will vigorously wag her tail—called flagging, she may have some discharge, and she will cry a lot. She may try to mount other goats, or vice versa. If there are bucks around, she will go stand at the fence line and stare at them or invite them over.

Goat keepers who don't have bucks sometimes have a harder time telling that a doe is in heat because the signs are more subtle. In this case, a "buck rag" often does the trick. A buck rag is a rag that has been rubbed on a stinky buck and then kept in a closed container. It can be brought out for the doe to sniff every day, or when the owner suspects the doe is in heat. If she is in heat, she will get excited by the smell and start to exhibit the signs discussed above. Goat keepers who keep a wether (castrated male) have also found that he is a good barometer of heat, acting "buckish" around this time.

> **Trick:** If you have no bucks or wethers, you can use a "buck rag" to tell when does are in heat.

The period of gestation—the time between breeding and birth—in a goat is around 150 days, or about five months. For the first three or four of these, the bred doe will not look any different. That last month is the most important in terms of growth of the kids, so the doe's nutrition and food intake needs to change to reflect that and to prevent metabolic imbalance at this critical time.

Pre-breeding
by Michelle Konnersman, DVM

Abortion Prevention Planning

Decide whether an anti-abortion vaccine will help your herd. If you have had problems with weak kids (some people call these "preemies"), dead kids or early or late term abortions last breeding season, an available vaccine may be helpful. If samples of the dead fetuses or kids were sent to the lab so that a diagnosis could be made, you will have a better idea whether to practice this prevention. A Leptospirosis 5-way vaccine may help to prevent up to 20% of abortions in goats. It is fairly inexpensive and easy to give. I believe it is good insurance against some abortion in goats. The Lepto vaccine should be given before breeding. Two injections are necessary, given 4–6 weeks apart.

If *Chlamydia* is the cause of abortion in your herd, pre-breeding is the time to begin vaccination against this cause of abortion. You can purchase a *Chlamydia psittaci* bacterin which should be given 60 days before breeding and repeated 30 days later. This vaccine can cause soreness and put the goats off feed for a day, so use it only if you know that it is needed in a herd.

Toxoplasmosis can be prevented, not by vaccine but by adding Deccox to the feed at a certain level. Speak to your veterinarian and feed supplier to determine whether you can plan ahead on feeding the Deccox feed. Deccox feeding should start by day 45 or 50 of the pregnancy to be really effective. Many *Toxoplasma* infections occur early in gestation and then the kids are infected, die and abort fairly early compared to *Chlamydia*.

Pre-breeding Supplementation

If you are in a selenium-deficient area, consider giving all breeding stock a selenium/vitamin E booster at least 30 days before breeding. It helps buck and doe fertility. An injectable supplement called Bo-Se is available. Oral pastes are now available for those goat keepers who do not like to give injections. The paste can then be used again during pregnancy. Current thinking is that Bo-Se injections not be given in late pregnancy because some goats and sheep may abort. (See Medications for more on Bo-Se.)

Pre-breeding Deworming

Plan to deworm breeding does before they are bred. During the first 25–30 days of pregnancy in a goat, the fetal cells are rapidly dividing. Giving any drugs during this period can cause birth defects in the fetus. Dewormers that are known to cause birth defects in goat kids are Valbazen (albendazole), oxfendazole and cambendazole.

Caprine Arthritis Encephalitis Virus (CAEV) Prevention

CAEV prevention is a lot easier if only a few goats need the special attention. Breed to CAE negative bucks only or use frozen semen. Make sure you know the expected kidding date and attend to the valuable does that remain in a CAEV-positive herd during the birth of their kids.

From "Ask the Vet," *Ruminations* #50

CDT Vaccination

Give does who have not been previously vaccinated a CDT vaccination 60 days and 30 days prior to the expected kidding date. Give a CDT booster to does who have been vaccinated 30 days prior to kidding. This will ensure that kids have immunity (through the colostrum they receive after birth) for the early part of their lives.

Bucks

If you are a new goat owner, don't be surprised if your bucks start peeing on themselves, fighting, mounting each other (and the livestock guardian dog) and acting generally disgusting. The urine is perfume to the does, though.

- Deworm your buck(s) and give a supplement Bo-Se shot at the beginning of breeding season. This is also a time to trim hooves and do a general check of the buck's condition.

- Consider providing some supplemental grain and warm water to bucks during breeding season; sometimes they are so worked up they forget to eat, so they need the extra energy.

- Some bucks will get urine scald on the muzzle or, less frequently, other parts of the body. A coating of petroleum jelly, herbal ointment or zinc oxide will help healing and provide a barrier to further urine.

- In some cases, bucks will get overly aggressive with other bucks, or even the owner, during breeding season, even injuring them. Keep a close eye on them during breeding season and separate them, if necessary.

Inbreeding—Good or Bad?

by Cheryl K. Smith

Ask a number of breeders what they think about inbreeding of their animals and you will probably get as many answers. The basic idea of inbreeding is that of mating between biological relatives. Two individuals are related if one or more ancestors are shared by both individuals. Because of common ancestors, these two different individuals could share genes at a locus that are identical copies of a single ancestral gene.

Breeders of purebred livestock use the term "linebreeding," to cover milder forms of inbreeding. The difference between linebreeding and inbreeding is often defined differently for each species and often for each breed within the species. Inbreeding, at its most restrictive, refers to what would be considered incest in human beings, i.e., parent to offspring or full sibling to sibling. Uncle-niece, aunt-nephew, half sibling matings, and first cousin matings are called inbreeding by some people and linebreeding by others.

What does inbreeding (in the genetic sense) do? Basically, it increases the probability that the two copies of any given gene will be identical and derived from the same ancestor. In technical terms, the animal is homozygous for that gene. The heterozygous animal has some differences in the two copies of the gene. (Remember that each animal has two copies of any given gene—one from the father and one from the mother.) If the father and mother are related, the two genes in the offspring may be identical copies contributed by the common ancestor. This is neither good nor bad in itself. It only becomes bad when the duplicate genes being expressed are those that cause a problem. For example, while cousin matings in humans represent only 0.05% of matings in US whites, 18–24% of albinism (lack of pigment) and 27–53% of Tay-Sachs cases in US whites come from cousin matings. This same is true for other recessive genetic diseases.

Despite the fact that inbreeding is usually equated with unhealthy offspring, and in dairy cattle, a decrease in milk production, some breeders have had good success with it as a breeding strategy. Now, a new gene study backs that up, suggesting that inbreeding can also create healthy new breeds—in the long run. A Scottish research team examined the DNA of the Chillingham cattle and found it "almost genetically uniform."

These fierce white beasts have lived wild within the walls of a forested game preserve at Chillingham Castle in northern England for hundreds of years. The Chillingham Wild Cattle are one of the original herds of wild cattle that still roam in their natural surroundings over about 300 acres of Chillingham Park. Many of the behavioral traits of these animals can still be seen in domesticated herds of White Park Cattle.

Though their origin is uncertain, the existing herd is thought to have been at Chillingham for at least the past 700 years. Prior to that, they likely roamed the great forest that extended from the North Sea coast to the Clyde estuary; and presumably when, some time in the 13th century, the King of England gave permission for Chillingham Castle to be "castellated and crenolated" and for a park wall to be built, the herd was corralled for purposes of food.

The shape of the skull and the manner in which the horns grow out from it are similar to the Aurochs (bos primogenius) and quite different from the skull of the Roman imports (bos longifrons). Because of this, many people believe that the Chillingham Wild Cattle are the direct descendants of the original ox, which roamed the British islands before the dawn of history. How they came to be white is another interesting matter for speculation. They invariably breed true to type and have never been known to throw a colored, or even partly colored, calf.

In recent years, blood samples from some of the cattle were analyzed from the genetic point of view and the blood grouping was found to be unique among western European cattle. So their origin still remains a mystery.

For the past 700 years they have been inbreeding and, as far as can be determined, the only effect has been that they are now somewhat smaller than they used to be. Their remarkable survival may be due to the fact that the fittest and strongest bull becomes "King" and the leader of the herd. He remains King for just as long as no other bull can successfully challenge him in combat, and during his period of kingship, he will sire all the calves that are born. Nature seems thus to have ensured the carrying forward of only the best available blood.

Over the years, Darwinian natural selection may have purged from the herd's DNA the dangerous recessive genes that normally make inbreeding risky. Chillingham calves unlucky enough to inherit two copies of a lethal recessive gene generally died before they could pass on their genes.

When animal keepers create a new breed of domestic animal, they use inbreeding to reinforce a desired trait. The animals often may have genetic problems, but by the seventh generation, the survivors are as vital and fertile as the original population.

Goat breeders who choose to use line- or inbreeding as a strategy for herd improvement should make sure they have a clear understanding of genetics, and be willing to lose some animals to genetically-caused health problems. But keeping accurate pedigrees and records of results may eventually pay off in superior animals.

From *Ruminations* #30

Why AI?
by Stacy Morris

Is your breeding program living up to your expectations? Do you ever get the feeling that you should be making more rapid and consistent genetic improvement than you are actually seeing? Is the hassle and expense of maintaining a pen full of bucks getting you down? If you answer yes to most of the above, you should give serious consideration to the use of artificial insemination (AI) in your herd. It will open up a whole new world of breeding possibilities to you, and can give you a big jump on genetic improvement within your herd. Getting started in AI is easier and cheaper than you might imagine. With a little practice, inseminating your doe can be faster and easier than breeding to a live sire (especially if the doe has taken a liking to the wrong buck!).

How difficult is AI?

Not very difficult at all. Taking an AI class or having someone experienced show you is the best way, but learning from a video or book is certainly possible. The first step is to look at all your does with a speculum prior to breeding them. It will not only help you become familiar with the process but can help determine whether the doe has an infection or even lacks a cervix. Like most things, practice makes perfect.

What equipment will I need?

The first thing you need is a liquid nitrogen tank or easy access to one. Tanks run about $250–500 used and $500–1000 new, depending on the size and holding time you select. Then you will need a good thermometer, an inseminating gun and sheaths, several sizes of speculums, scissors or a Cito cutter, a thaw jar, non-spermicidal lubricant and a light source. Depending on how fancy you get the cost will be between $100–150.

Finally you will need semen—typically it runs around $25 per straw, but can be higher if the buck is well known and/or dead, or less if the buck is unknown and unproven. The bottom line is that getting set up with AI equipment can be as expensive as purchasing a very well bred buckling, but the tanks do not eat as much and they allow you access to many different sires/bloodlines—something not possible with live breeding.

Can I breed all my does AI?

In theory you can, but realistically you will have better results using AI only on those does that have well-defined heat cycles lasting no more than 48 hours and with no history of breeding or kidding problems. Doe kids often are very difficult to catch in strong heats and may physically be too small for successful inseminating. Many people I know who do a lot of AI, breed all of their adult does AI once or twice and then use a live buck if they have not settled via AI. Having a buck available for help in heat detection is also helpful. The success of your AI program relies heavily on accurate heat detection and proper timing of the insemination.

Does AI have health advantages?

Yes. AI is inherently more hygienic than natural service, because it eliminates physical contact. Many herds are closed to outside breeding and AI may be the only way to use a sire from a closed herd, or to add outside bloodlines if your herd is closed. The incidence of disease transmission attributable to AI is extremely rare.

Semen

Semen has shown to be viable for up to 40 years and possibly longer, a fact learned from tests done in bull facilities. There is no reason to think goat semen would not be viable for at least as long, and I personally have used 20+-year-old semen with good success. Proper storage and handling is absolutely essential, however.

> **Tip:** [A] buck being collected needs to be in good physical condition, with recent selenium, vitamins and grain to boost his condition. He should be current on his vaccinations, have his feet and belly trimmed, not be stressed, and should be used a week or two before collection to clean out old semen. He also should have had a break from breeding of between four days and a week.
>
> From "Karin Reyna: Semen Collector," *Ruminations* #46

Every time the temperature goes up and down, sperm cells die. An accurate thermometer for thawing straws is a necessity. Once semen is thawed, keep it at a stable temperature and get it in the doe in a minimum amount of time. Ultraviolet lights (fluorescent) and rough handling (dropping or thumping the straw) will also kill semen.

Good Records

We learn by mistakes, and keeping good records on each insemination will help you on future AIs. A good record describes the length of the doe's heat, the stage of heat at insemination, amount and color of mucus, location of the cervix and the amount of penetration with the gun. If you have a microscope available, look at a drop of semen to make sure it contains live sperm. This usually isn't a problem, but if the semen is old, verifying that it is still viable is wise. Good records may be the key to better conception rates!

From *Ruminations* #51

Hypocalcemia: How to Recognize, Treat and Prevent It

by Sue Reith

Symptoms

Hypocalcemia symptoms don't appear until after the third month of pregnancy or later. The first sign that your doe is headed in that direction will be that she loses interest in her grain ration. This could take place suddenly, or it might happen more gradually. Shortly thereafter she will lose interest in her hay as well. Often this doe has been getting a regular, hefty ration of grain along with her hay during the early months of the pregnancy. And there is an even greater chance of this happening if the doe was getting grass hay instead of alfalfa along with her grain ration during that period. If you are sufficiently concerned to take her temperature when this first happens, it will be normal (102.3° F.), but will drop to sub-normal (below 102° F.) as the signs progress.

If you fail to take immediate corrective action, you can expect her to weaken fast and start acting lethargic and depressed. Her rear legs will appear wobbly. If you don't intervene with corrective measures right away, she will go down and not get up. When I first went through this, back in the 1970s, I assumed that my doe just wasn't hungry, and that she was lying down because she was uncomfortable from lugging around all those babies inside of her. It took me a day or two to start being concerned. I took her to a dog/cat vet who didn't recognize what was wrong, so he read his veterinary guide (which was, and still is, generally devoid of any sort of information about how to recognize and treat hypocalcemia), after which he decided it was pregnancy toxemia and put her on IV Ringers until she died.

For readers who don't know what hypocalcemia is, it's a serious condition where a pregnant or lactating doe isn't getting enough calcium to support herself and at the same time provide for her growing fetuses (or to produce milk if she is lactating). It's caused by a dietary imbalance that deprives her of the calcium she needs to support all of the work her body must do.

I find it really interesting that a dog/cat veterinarian caring for a client's breeding bitch or queen in the last stages of her pregnancy stresses the importance of high quality nutrition including plenty of calcium so as to prevent hypocalcemia as the patient nears term. But I have yet to see such concern for the quality of diet, or of calcium intake, of a pregnant doe. In fact, in a most bizarre reversal of that concept, I have observed that goat owners are routinely advised to reduce the quality of nutrition, including calcium, provided to their pregnant does in the last few weeks before they freshen. As a woman, I know that I would consider a suggestion from my obstetrician that I severely cut back on my nutritional intake, including calcium, in the last few weeks before I was due to deliver a baby, to be incredibly stupid.

What follows now is an overview of what you, the doe's owner, can do yourself to repair your pregnant doe if she starts showing those first signs of hypocalcemia. I will be mentioning dosages for meds that are intended for full-sized dairy-type does, weighing 120–150 lb on average. If the doe you're working with is a small breed you'll need to adjust the doses accordingly.

Treatment

Right away, and speed is important, start her on Nutridrench or oral propylene glycol to restore the energy she lost when she stopped eating. This will prevent ketosis, a metabolic problem caused by an animal's living on its own body reserves because it has stopped eating food. If you don't correct this fast she can sink into a coma and die. She should get the suggested amount of Nutridrench for her weight, or 60 ml of propylene glycol, 2X daily for two days to restore her, and then she should get 30 ml a day from then on until she is eating right again, to prevent/reverse ketosis, a life-threatening condition.

> **Tip:** When a pregnant doe shows signs of hypocalcemia—losing interest in grain, and then hay—immediately start her on Nutridrench or 60 ml oral propylene glycol twice a day to restore the energy she lost when she stopped eating.

Now let's turn quickly to the primary problem, hypocalcemia. Calcium is vital to muscle tone. The heart is a muscle, and for the heart to beat properly the doe must have sufficient calcium. So as soon as she's been treated to prevent ketosis, start her in right away on calcium replacement therapy to restore her calcium level to normal. Calcium gluconate is often used for this, and if you are bringing a vet in to help with the repair process he or she may recommend its use for your doe. But my own preference is a product called CMPK (or its generic), because calcium gluconate contains only calcium, but CMPK also contains magnesium, phosphorus, and potassium, all of which make the calcium available faster and more efficiently to the body and speed up her recovery rate. Additionally, there's such a thing as "too much of a good thing," and when calcium gluconate is used to treat hypocalcemia there is a grave risk that giving it too quickly or in too large a dose could cause her heart to beat too fast, putting her at risk for heart failure. But the potassium in the CMPK slows down the heart rate. So the calcium and the potassium balance each other out in CMPK to maintain a level heart rate as the calcium is restored in her system.

CMPK is dosed at the rate of 30 ml SQ about every two hours until the calcium she needs has been replaced. We dose in this manner so that we can continually monitor her progress to determine when her system has returned to normal. To do this, simply compare her heart rate to that of a normal doe (70–80 beats per minute). At the beginning of her treatment the hypocalcemic doe's heart rate will be slower than that of the normal doe because of the calcium deficiency. When her heart rate is the same as, or a tiny bit faster than, the normal doe's heart rate, and she appears bright and alert and wants to eat again, you will know things are going well.

Keep in mind that a single dose, or just a few doses, of this product will *only* balance the doe's calcium level *for the moment*, but as those babies continue to grow they will drain more and more calcium from her, so after you bring her heart rate back up to normal it's crucial to continue giving her *daily maintenance doses of about 30 ml (1 oz) until she freshens. If you notice her weakening at any time before then you w*ill need to increase the dose again, but only temporarily, until she returns to normal.

If possible, it would be better to use injectable CMPK instead of the oral form to treat your doe. This is because: a) It's always risky to dose a very weak animal orally because part of the fluids can end up in the lungs if she struggles, adding aspiration pneumonia to her problems, and b) calcium is, in concentrated form, quite irritating to the tissue, and it can burn the tender membranes of your goat's throat. The problem with my telling you this is that injectable CMPK, while relatively inexpensive (about $4/1000 ml in the catalogs) is only available by veterinary prescription. The down side of that is that some veterinarians, especially those that are not goat-oriented, may not know about hypocalcemia, or why continued doses of calcium are necessary. So they could be overly-cautious about your using it, only prescribing/providing a single dose, or perhaps two. This is like trying to fix a leaking dam by putting your finger in the hole. All the more reason why you might want to share this article with the veterinarian of your choice.

Once her calcium level has been regulated and she starts eating again, she will probably refuse grain for a while. Don't worry about that because her instinct is still trying to regulate her calcium-deficient condition and she's the best monitor of that. Pretty soon she will probably start eating the grain again, at which time she should get just a small amount at each feeding.

If she's not eager at first to eat her hay again (grass is okay to start with if that's what you have available, but alfalfa or alfalfa pellets would be a really good choice now) it would be a good idea to bring her some of her favorite browse. (I feed salal and wild huckleberry up here in the Northwest, both of which stay green all winter.) In your area there must be something yummy that is, of course, not toxic. If you don't know her favorite, give her a variety and let her choose.

Please keep something else in mind while treating your hypocalcemic doe: If she has been down for three or more days in this weakened condition, it's important for you get her back up on her feet as soon as possible. Otherwise her legs will lose their muscle tone fast and won't be able to support the heavy weight of her body if she tries to get up on her own. And additionally, if you let her stay down for too long a time her leg joints may begin to ankylose (freeze permanently in the bent position). This is irreversible. To prevent it you may have to create a makeshift sling, attaching it to a pulley that is fastened to an overhead beam in the barn. About every two hours the sling should be raised up so that she can touch the ground comfortably with her feet and move around, and then lowered again so she can rest for a while... The process should be repeated continuously, two hours up and two hours down, until she can support her body's weight with her own legs again. This usually takes only a few days, but in her pregnant condition she might have a harder time getting her strength back. If you want a picture of a sling I have one in my archives, write me at suereith@msn.com.

Prevention

You probably are curious as to how your goat got hypocalcemia in the first place, and how you can keep her from getting it again next time... Well, it has to do with what you were feeding her before, during and right after she was bred. If your doe was either under a year old or still milking when she was bred, she should continue getting the ration you had been giving her to support her growth or milk production. But if she was an adult, and 'dry' (not lactating) when you bred her, she should have been getting little or no grain at that time, and for the first three months of her pregnancy, because her body isn't yet supporting rapid fetal growth. You see, at three months the fetus is no bigger than a newborn baby kitten.

But once that first 90 days or so have passed, the babies are completely formed and start to grow rapidly. As they continue to grow over the next eight weeks they will need more and more calcium. So at that three month point you should begin giving a small amount of alfalfa with the grass hay, increasing it gradually until at the time of freshening she is getting all alfalfa, which in my view should be continued for as long as she's milking. The alfalfa should be tapered off as her milk production tapers off, until she reaches the end of that lactation, at which time she can go back to grass hay.

At that same 90-day point when you start giving the pregnant doe alfalfa, you should begin offering a handful of grain at each feeding. That amount should be increased slowly over the next 60 days so that by the time she freshens she is getting maybe a cup in the morning and a cup at night. Then, depending on the amount of milk she gives per milking, you should increase the grain so that she is getting enough to help produce the milk but not make her fat. A pound of grain is usually recommended for 8 lbs (about a gallon) of milk. I add alfalfa pellets to a doe's grain to keep her busy while I milk her out.

Cause

This hypocalcemia (calcium deficiency) problem that can show up anytime in the last 6–8 weeks of the doe's pregnancy is pretty simple to explain. It has to do with ratios. Most of us know that a ratio of 2 parts hydrogen to 1 part oxygen (H_2:0) makes water. If you don't have that ratio, you don't have water. And those among us that are raising goats in copper-deficient areas might be aware that there has to be a ratio of 10 parts copper to one part molybdenum (10:1) in the soil for copper to be available for our goats. Well, a ratio of 2 parts calcium to 1 part phosphorus (2:1) is needed to make calcium available to us, AND to our pregnant/lactating goats. So you can see now how important it is that all of these ratios stay balanced. When they don't, the substance we need will not be available to us. So… the goat needs to be regularly eating a ratio of at least 2 parts calcium (abundant in alfalfa) to 1 part phosphorus (abundant in grain) to make the calcium available to her body that she needs to support both herself and the babies that are growing rapidly in her uterus (or for lactation if she is in milk). If her owners do not provide the correct 2:1 balance for her in the feed she is given, she will become hypocalcemic.

As noted before, at the beginning of the pregnancy, with the exception of yearlings that still need calcium for growth and does that are still lactating, a doe should be getting either just grass hay with no grain, or some alfalfa with a tiny bit of grain, while the fetuses are in that early stage of development. Unfortunately, many new goat owners (including me, early on) think that they should feed the doe lots of nutritious grain from the moment she is pregnant to assure that the babies will grow big and healthy. This is the first step towards hypocalcemia. At the three months' point when the babies do start growing rapidly, the doe suddenly starts needing more calcium to support their needs.

If she has been given lots of grain in those first three months, her amazing instinct tells her to stop eating that grain to free up calcium from the hay. It's even more problematic for her if the grain she was getting early on was combined with grass hay, which has hardly any calcium in it. Tuning in to her body, she stops eating that large amount of high-energy grain it's become accustomed to, and as a result rapidly becomes nutritionally deprived, on top of the calcium deficiency she's already experiencing. She weakens fast, becomes lethargic and wobbly, and goes down, and owners and consulting vets stand around scratching their heads, not knowing what led up to this.

They (might) offer her Nutridrench or propylene glycol to correct the ketosis she's headed for (or already in), and then they wonder why she continues to get weaker and weaker, not realizing that the mismanagement of her feeding program has put her in a state of hypocalcemia. In addressing just the second part of this twofold problem (hypocalcemia leading to ketosis), they fail to provide a calcium replacement, vital for her own muscle tone as well as for fetal development. No calcium, no muscle tone, no heart pumping, dead goat.

Those of you who read the various popular veterinary guidebooks have probably noticed that they mention little or nothing about hypocalcemia, which, by the way, is a correctable metabolic disorder (condition), and not a disease. You can be sure that I, too, have noticed the lack of coverage of the subject in those same books, which is why I have written this article. I have been making a valiant effort to encourage the veterinary authors of these books to update their work to include it, but progress is slow. While most goat management coverage in our current resource books is quite helpful, in this particular area I see the following deficiencies:

In the *Merck Veterinary Manual* a single sentence addresses hypocalcemia. In the 8th edition on page 744 (the "Pregnancy Toxemia in Ewes" section), under "Diagnosis," the single sentence reads: "Hypocalcemia, uncomplicated by pregnancy toxemia, should always be considered for recumbent late-gestation sheep." That's all that is said! There is no discussion anywhere in the book of either how to recognize it or how to treat it. I can only speculate that this is because the condition is not well understood by the author.

Another popular goat management reference book, Smith and Sherman's *Goat Medicine,* does a good job in the section on goats' dietary needs of pointing out the importance of a diet containing 2 parts calcium for every 1 part phosphorus. Unfortunately, that information was not carried over into other areas of the book where it is needed. For example, in a discussion of 'metabolic disorders appearing in late gestation' a reference is made to hypocalcemia, which is then followed up by a comment about a chemical imbalance within the hypocalcemic goat that makes calcium unavailable to her. The author misses this golden opportunity to explain to the reader (who often does not know the nutritional needs of a pregnant/ lactating goat) the importance of providing a diet that contains 2 parts calcium for every 1 part phosphorus to free up calcium for her fetuses/milk production to prevent hypocalcemia. Once again, I am left to assume that the authors are not making this connection? Oddly, there is another comment in that same section on metabolic disorders about goats needing "2 parts 'forage' to 1 part 'concentrate,'" a misleading statement at best, and a recipe for disaster at worst, since vast numbers of goat owners have only grass for forage, and grass contains very little calcium at all.

John Matthews, in *Diseases of the Goat,* talks about "hypocalcaemia," noting that it may appear in late pregnancy… AND in any stage of lactation, an important bit of information! But he, too, misses the simple cause, a dietary imbalance that prevents the uptake of calcium from the feed. He relies instead on an unexplained "failure in the homeostatic mechanisms to meet the increased demand for calcium." However, in his discussion of hypocalcaemia, he redeems himself admirably by saying that: "All recumbent or comatose goats should be treated as potentially hypocalcaemic and given calcium."

With the exception of John Matthews' book, none of the available goat management references that I have seen place any emphasis on hypocalcemia whatsoever, despite the fact that it's probably the most common reason for pregnant/lactating does managed by inexperienced goat owners to "go down." As a result, it is entirely overlooked by veterinarians who use these books as guides when they attempt to come up with a diagnosis and treatment for the animal. The huge down side of this is that when the hypocalcemic condition is not recognized, the veterinarian that is inexperienced in diagnosing "down" pregnant goats will, using these books for reference, almost always opt for a diagnosis of pregnancy toxemia, pregnancy ketosis, or any combination or variation of those words.

When asked by the client about the possibility of this being a calcium deficiency, the vet commonly responds, "This goat's problem has nothing to do with calcium." The predictable treatment regimen will then be: "Treat with propylene glycol for ketosis, and get the babies out fast to save the life of the doe." To do that he generally suggests a c-section or, even worse, a shot of Lutalyse to abort the doe, which doesn't work because without calcium the doe goes into labor but the uterus has no muscle tone to expel the fetuses.

In providing the "down" doe with calcium replacement therapy instead of taking such drastic measures, I have never experienced loss of either the doe or the kids, and does treated to reverse the hypocalcemia condition routinely freshen normally. If my goat had this problem and I sought the help of a veterinarian, I would want that vet to become my partner in undertaking this treatment before even considering anything so drastic as c-section or abortion!

Many well-respected dairy goat nutritionists and veterinarians have mentioned hypocalcemia in their work over the years, but most have never dwelled upon the cause of it, which is vital to its treatment. Since livestock management is not their chosen field, maybe they assume that all dairy goat people instinctively know how to feed their pregnant stock correctly?

Following are a couple of individuals whose contributions have helped us to understand goats' dietary needs as well as potential corrective measures to take when the needs are not met:

Dr. M.E. Ensminger, a well-known livestock nutritionist whose work has guided many of our experts today as they figure out what to put in livestock feeds, says in his "bible" of livestock nutrition, *Feeds & Nutrition—Complete*, published in 1978, that alfalfa (lucerne), a legume, "is high in calcium, protein, and carotene, and in many other minerals and vitamins." He notes further that "legumes are excellent calcium sources, while grasses and silages tend to be substantially lower in calcium content." He points out that both bone growth and lactation (and muscle tone as well, by the way) require substantial quantities of these minerals. He says, "If there is a severe imbalance of them during pregnancy and early lactation, 'milk fever' may occur." He continues, "Therefore, in order to prevent these problems, the ratio of calcium:phosphorus should be at least 2:1." (2 parts calcium: 1 part phosphorus.) (He also states: "In males an imbalance of calcium to phosphorus often leads to the development of urinary calculi.") Finally, for those of us who rely on grass hay to feed our goats, Dr. Ensminger says that where additional calcium is needed, ground limestone is usually the mineral of choice, but if the animals are in need of both calcium and phosphorus the best choice for provision of these two essential minerals would be dicalcium phosphate (2 parts calcium: 1 part phosphorus).

As an addendum here, in further support of Dr. Ensminger's findings, books providing nutritional guidelines for humans note in mineral requirement charts that a symptom of excess phosphorus intake is "decreased blood calcium." A symptom of deficiency in phosphorus intake is "general weakness." This completely supports Dr. Ensminger's findings that a severe imbalance in either direction of the 2 parts calcium to 1 part phosphorus requirement in the diet makes calcium unavailable, the result being lack of muscle tone (i.e., general weakness).

Another knowledgeable person I had the good fortune to come into contact with in southern California where I first began to raise dairy goats was a veterinarian named Dr. Robert A. Jackson. He was what you could call a goat vet's goat vet, and he and a dairy goat breeder/judge named Alice Gaye Hall often co-wrote dairy goat management articles. In one of them, printed in the July '82 *Dairy Goat Guide* and called "What to Know about Medications," he told the readers that it's important for goat owners to keep calcium in their cupboards because goats often come down with "eclampsia, which is much like milk fever...." Just as others do, they called it milk fever. And while they don't spell out why it happens, they do make the observation that a calcium deficiency (hypocalcemia) sometimes exists in the pregnant/lactating goat, and that the owner should be prepared to treat it when it shows up.

From *Ruminations* #41

Ketosis: What Is It and How Does It Happen?

by Sue Reith

Ketosis is a word that gets bandied about a lot, indiscriminately in my view, and it is incorrectly said to be the primary cause of a number of ailments. One that comes to mind, because I see it mentioned often, is "pregnancy ketosis." There is no such thing! There is pregnancy, which is one condition, and there is ketosis, which is an entirely separate condition. Ketosis can develop as a secondary condition under some circumstances during pregnancy or lactation, but it isn't limited to pregnancy or to lactation, and it can show up at any stage in the life of a goat. (By the way, ketosis happens to people, too.)

The word itself is not well understood, so I'll try to explain it. In technical terms, it's a condition brought on by a metabolic imbalance. In scientific terms it's defined as an accumulation of excessive amounts of ketone bodies in body tissues and fluids. "Ketone bodies" are the metabolic substances known as acetoacetic acid and beta-hydroxybutyric acid. Acetone, which puts off the peculiar odor associated with ketosis, comes from acetoacetic acid. These substances are all normal metabolic products of 'lipid' within the liver. When they become severely imbalanced as the result of ketosis, the liver will fail.

Cause

By its very nature, ketosis has to be a secondary condition, because it's the direct result of a process that starts when the animal, for whatever reason, stops eating. Why the animal stopped eating is the primary factor that must be addressed and corrected. And that must be dealt with quickly, because when it stops eating, the lack of an external energy source forces it to use its own reserves to provide energy. These reserves are in the form of fatty tissue.

In the words of W.C. Allenstein, DVM, a cow vet who wrote for *Hoard's Dairyman* for many years, "When this fat utilization occurs, free fatty acids are released into the blood stream and are used by the liver for energy. If this occurs at too fast a rate, the liver is bombarded with too many fatty acids, and there is an increase in ketone bodies released into the system. At a certain level the classic symptom of acetone odor on the breath and in milk [if the animal is lactating] will occur... The ketone bodies formed by incomplete fat metabolism by the liver create these symptoms." He goes on to say: "Today we know that anything that disturbs the body—other diseases, missed feedings, conditions disturbing feed intake, will create ketosis."

Lori Ward, then a student in Dairy Sciences at the University of Wisconsin, noted in an article that when the animal is forced, by lack of an external energy source, to turn to its own body for sustenance, "The body fat is mobilized to supply needed energy. The mobilized fat is processed in the liver, and it tends to accumulate. In most fatal cases the postmortem findings reveal a fatty liver. During fat mobilization, ketone bodies (one of which produces acetone) are produced and circulated in the blood, hence the names ketosis or acetonemia." Lori notes that most of the accepted ketosis treatments attempt to raise blood glucose in some manner. This provides a quick energy source for the victim, ending its need to live on its own fat reserves to survive.

A classic example of how ketosis gets involved (and often is the only disorder that gets recognized and treated, the end result being the loss of the victim's life) is found in the prior article I wrote on hypocalcemia. When a pregnant doe is being fed a dangerously imbalanced ration and stops eating a large part of that ration to instinctively try to correct the imbalance, the loss of this external energy source makes her turn to her own body's resources for survival. Since the babies are still growing in her it is very important to fix the original nutritional imbalance, which in this case is a diet-induced calcium deficiency. The fetuses inside of her are draining her of her own calcium, because of which she loses her muscle tone. Without calcium she becomes very weak. So she has weakened muscles, including the heart muscle, and at the same time is living on her own body reserves because she has stopped eating (and soon is too weak to eat) her imbalanced ration.

When a pregnant doe becomes hypocalcemic and is misdiagnosed by a vet who doesn't understand the dynamics involved, the vet fails to give her calcium replacement therapy at the time he or she is reversing the ketosis with energy replacement substances such as propylene glycol or Nutridrench. The result is a drained and weakened system despite the treatment for ketosis, and she inevitably dies either of "unexplained" causes, or of what the vet labels "milk fever," "pregnancy toxemia," "pregnancy ketosis," or "parturient paresis," all of which are misnomers. Then if he does a necropsy he will generally label the cause of death as "liver failure." I don't agree that liver failure is the actual *cause* of death. I see it as the *result* of the animal's not having been diagnosed and treated for both the primary cause, hypocalcemia, and the secondary cause, ketosis, because of which she dies. The liver failure, then, is not the cause of death, but simply the end result. Ketosis is the secondary condition involved. The survival of the animal is dependent upon the discovery and correction of the primary condition.

From *Ruminations* #41

Abortion in Goats of North America

by Cheryl K. Smith

We breed our goats for a variety of reasons, my favorite being to make milk (although I love those kids). But breeding any animal creates a risk that the animal's health may be endangered or compromised in some way. Fortunately, goats generally get through the pregnancy in good health and have no problems kidding.

Despite good management, however, there may be a time when you are faced with kidding problems, including abortion. Abortion is the termination of pregnancy after the animal is formed but before it can survive. If it is born at full term, but dead, it is called a stillbirth.

Because of the potential for abortion in a goat, understanding the potential causes and preventing them is important.

What causes abortion in goats?

When a doe aborts early in her pregnancy, it may be hard to determine definitively that it was not actually a failure to conceive or a false pregnancy. You may not even notice that something happened. Early abortions are most often due to malformations in the fetus, so they are not viable. At times, infectious diseases can cause early loss of kids, depending on when the dam was infected.

Another case where abortion is likely is when a male sheep breeds a female goat. Normally, in this unusual case, the embryos are usually not able to survive past the second month of gestation. In very few cases have such hybrids survived to birth.

Other causes of abortion in goats include stress, infection and poisoning. However, even with laboratory testing, the cause of as many as 70% of abortions cannot be identified. This is because the fetus may have died weeks or months prior to the actual abortion, genetic or toxic causes are not identifiable in the samples submitted, the fetus may be mummified, or the cause is unknown.

STRESS. Stresses that can lead to abortion include cold weather, poor feed, overcrowding, unclean conditions, pregnancy toxemia, and vitamin/mineral deficiencies. Proper shelter, a well-balanced diet with mineral supplements, and enough space are the essentials. Do not overfeed or underfeed, and increase the ration during the last eight weeks of pregnancy, for optimum growth of the kid.

In some cases, goats have been known to abort their kids as a result of trauma—usually after getting butted by another goat.

INFECTION. Fifty percent of all abortions in goats are considered to be infectious in origin, so this is the first place to start in determining the cause. A number of disease agents are responsible for causing abortive processes in goats.

Chlamydiosis/Chlamydiasis. *Chlamydia psitacii* is responsible for many caprine abortions in North America. It was first reported in Germany in 1959 as a cause of caprine abortions.

This bacterium causes abortion in the last trimester of the pregnancy, particularly the last two weeks, as well as causing pneumonia, lameness and pinkeye in the dam. It is also contagious to humans, causing a flu-like illness.

The bacteria get into the intestinal tract of a pregnant doe and, once it is in the blood, it reaches the fetus through the placenta. The fetus dies and then is aborted. The fetuses are often mummified.

Commercial vaccines are available, which help prevent abortion, but not the infection. Once the disease is diagnosed, treatment consists of a course of LA-200 or Tylosin.

Brucellosis. This disease is caused by a microscopic organism found mainly in cattle. It causes late abortion, infertility, swelling of testicles, fever and other infections. Kids that survive are often very weak at birth. The disease is transmitted through milk or vaginal fluids. A vaccine is available, but it can cause abortion. Many states are now certified brucellosis-free, so this may not be a likely cause of abortion in most goat herds.

Campylobacteriosis. This disease is orally transmitted, and may cause up to 70% of the does on a farm to abort. Abortion occurs in the last six weeks of pregnancy.

Cache Valley Virus. This virus affects sheep in some parts of the US; it is not yet proven to affect goats. It causes fetal abnormalities which may lead to abortion. It is seen when the first trimester of pregnancy occurs during biting insect activity. The dam will have antibodies at kidding, even if she is no longer affected.

Border Disease. This infection is caused by a pestivirus that is related to bovine viral diarrhea (BVD). Abortion caused by border disease may occur at any time during the pregnancy and the fetus is often mummified. Kids that are liveborn are weak.

Leptospirosis. This disease can cause abortion in goats, although the actual frequency is unknown. Goats can get it from an environment contaminated by urine from other infected animals. Symptoms include weakness, loss of appetite, and death.

Listeriosis. *Listeria monocytogenes* is the cause of this disease, which can lead to abortion in the last trimester. It can be spread in contaminated soil or poor quality silage. It causes fever, loss of appetite and decreased milk production in goats, as well as inflammation of the placenta, blood infection and eventually death of the fetus. This disease can be prevented by pasteurizing milk and feeding good quality feed to goats. *Note that recent cases of listeriosis in the U.S. have been caused by cheese made from pasteurized milk.

Salmonellosis. A variety of salmonella species cause abortion in goats. The most common one causes mid- to late-term abortions that are secondary to diarrhea in the goat. Other signs are fever, inflammation of the uterus and infected vaginal discharge.

Toxoplasmosis. *Toxplasma gondii,* a microscopic parasite, is the most common cause of abortion in North American goats. It occurs in the first half of pregnancy, although sometimes the fetuses are reabsorbed. The doe shows no symptoms and breeders are generally unaware that the pregnancy has been lost.

Many of us know about this because of warnings to avoid eating undercooked meat and coming into contact with cat feces during pregnancy. The same advice is given to people with weakened immune systems.

In goats, toxoplasmosis is often spread by cat feces in the pasture or in hay storage areas. This is a long-lasting germ, living as long as 18 months in soil. Young cats are more likely to spread it than old ones, so spaying and neutering your barn cats is essential.

Q-fever. The Queensland (Q) virus is caused by a bacterium called *Coxiella burnetti.* It is spread by tick bites, contaminated pasture and inhaling dust that contains it. Abortions occur at term (stillbirth), while some kids are born weak. The does have no symptoms. It is often hard to diagnose based on post-abortion lab tests.

Other infectious diseases that can lead to abortion include caprine herpesvirus, yersiniosis bacteria, and tick borne fever.

POISONING. A number of medications, including dewormers, steroids, and phenylbutazone, can cause abortions in goats. Levamisole, a worming medicine, is well-known for causing late term abortions when given to pregnant does.

Nitrates not only can poison goats, but can cause abortion with a less than lethal dose during any stage of pregnancy. Goats can get an overdose by eating pasture that has absorbed large amounts of nitrate from fertilizer or from plants, for example, kale or sorghum, that contain high concentrations.

Certain plants are known to cause abortion in goats for a variety of reasons. Some have alkaloids that accumulate if too much is eaten, some absorb selenium, leading to overdose, and some have unknown toxins. For this reason, it is important to know the plants that grow in areas frequented by your goats.

Many believe that pine or juniper needles can cause abortion; but according to *A Guide to Plant Poisoning of Animals in North America,* goats are not among the animals affected by them. Likewise, hairy vetch, which causes abortion in horses does not affect goats, or they don't eat it.

Plants containing dangerous alkaloids include locoweed, comfrey and tansy ragwort. With tansy ragwort, or *Senecio*, only 1% per day eaten by a goat can lead to abortion.

Plants that absorb selenium can often be found in areas known to have high concentrations of the element. These include milk vetch, snakeweed, woody aster, goldenweed, gumweed, and Indian paintbrush.

Other problematic plants include Hemlock and horsebrush, which grows with sagebrush mainly in the southwest US.

One good thing I have found is that my goats usually don't eat large concentrations of plants that are poisonous. I have never seen adverse effects, despite nibbles of fern, rhubarb and rhododendron. That is not to say, however, that goat breeders should not be aware of and attempt to remove plants that may endanger their goats.

What should you do if your goat aborts?

Remember that infectious disease is the most likely cause of an abortion, particularly if you have had more than one goat lose her kids. Sometimes what is known as an "abortion storm" occurs, where a large number of does on a farm have abortions. This points even more to a transmissible infection.

Keep detailed records of dates, symptoms, and other pertinent information. Always use gloves when dealing with the doe, the fetus and the placenta. If you are pregnant or have a weakened immune system, DO NOT be involved in the kidding process.

If you do not plan to submit the fetuses and placentas to a vet or lab for diagnosis, it is essential to properly dispose of them. Burning is recommended. If you do plan to have a necropsy or lab studies done, refrigeration is the recommended method for preservation. *Goat Medicine* suggests freezing the first placenta and fetus(es), while waiting to see if there is a herd-wide problem, although this is not ideal.

Isolate the doe, so that if she is infected, other goats are not put at risk. If the abortion occurs outdoors, *Goat Medicine* recommends burning the area using straw and diesel fuel.

> **Tip:** After an abortion, consider treating does that haven't kidded with 3 SQ injections of LA-200 at 9 mg/lb at three-day intervals.

The authors also suggest treating all does that haven't kidded, in case an infectious agent is implicated. They recommend three IM or SQ injections of LA-200 at 9 mg/lb at 3-day intervals.

Is it important to test to find out what went wrong?

Particularly if more than one goat is involved, testing is helpful and sometimes crucial to the health of the herd. Such a decision is individual and may depend on finances, proximity of a lab, or the ability of your vet. Remember to refrigerate the fetus and placenta.

Ideally, you should submit blood samples from the animals that aborted—10% of the herd in cases of abortion storms. A second sample 10–15 days later will help with diagnosis. Genital swabs from the aborting does can also help determine a diagnosis, and are important.

If the whole fetus cannot be sent, the most reliable alternative is its stomach contents. The placenta should always be sent, with the cotyledons intact.

If the fetus is mummified, most tests are not helpful, so unless a number of goats are involved, you may want to save the time, travel and expense.

The final thing that will assist in making a diagnosis is a full case history. This is where keeping good records is important. Was the dam sick at any time during the pregnancy? What were her symptoms? When was she sick? When was she due? If she had labor, did it seem normal? Was she treated at any time and with what?

Losing kids is a sad time for the goat and for us, the goat keepers. Being observant and having the information to know what steps to take and how to prevent other such losses is an important part of responsible goat keeping.

From *Ruminations* #37

How to "Read" Ligaments to Predict When a Doe Will Kid

Even very experienced goat keepers are occasionally caught off-guard by does who kid sooner or later than expected. No guaranteed method exists for determining exactly when a doe will kid, but learning to "read" the changes in a doe's pelvic ligaments is the most accurate predictor of impending labor.

The rump of a doe will normally be rather flat and solid. As she nears the end of her pregnancy, things will begin to change. Her tailbone starts to raise up and the ligaments that connect her tailbone to her pelvis get stretchy and loose. Her body is making a birth canal.

If you make a habit of feeling the ligaments for several weeks before she is due to kid, you can "teach" your fingers to know your doe's body. When pelvic ligaments "disappear" under your fingers, you can be fairly confident that she will kid within a day.

Here's the technique: Every day for a few weeks before she's due to kid, run your thumb and forefinger down the portion of the spine that runs from the hip bones (where the back and the rump meet) to the pin bones (the two bones to the side and slightly down from the tail base.

About two-thirds of the way down, you will feel two ligaments, one on each side, which come out at an angle. One helpful method is to close your eyes and imagine a peace sign, or a stick drawing of a bird's foot with three toes. The spine is the upright part and the ligaments make the two diagonal lines at the bottom.

As her due date approaches, everything will begin to feel mushy. The part of the spine you have been feeling each day begins to rise. As the ligaments lengthen, they begin to feel softer but you can still feel where they are. Eventually the tail head is so far up that a slight "hollow" forms between it and the body. When you can no longer feel the ligaments, she will probably kid within the next 24 hours.

From *Ruminations* #55

Kidding for Beginners
by Cheryl K. Smith

If you are a new goat owner, or have had your goats for a year or two and just bred them for the first time, you may be worrying about them getting through kidding safely. The main thing to remember that having a baby is perfectly normal, and most of the time, it will go just like nature planned.

Knowing what to do when a problem does occur is helpful, but don't anticipate problems. This article will give you an overview of what to look for in labor and how to recognize and deal with the most common minor problems that may occur.

I have found that Nigerian dwarves normally deliver their kids between 144 and 154 days. Most pregnancy calculators use 150 days for estimating kidding, so beware and keep a close eye on your doe from about 144 days.

How do you know what to look for?

The first thing is a softening of the tail ligaments. This is 100% effective, in my experience. If you check the doe long before she reaches this stage, you will know what you are feeling. I have been mistaken though, even recently, in a doe with very widely-spaced ligaments.

When the tail ligaments go completely mushy the doe will kid within 24, and often, 12 hours. This is the best sign that she is entering the first stage of labor. You can sometimes tell that this has occurred if the goat seems to have lost her ability to hold the tail up.

Also look at the udder. The udder will begin to develop and fill. In some does this will begin to occur months in advance, in others it happens in the last 3–4 weeks, and in other cases not until right before kidding. Checking regularly toward the end of pregnancy is helpful, if the doe is agreeable. Otherwise, just look at it closely for changes; it often becomes taut and shiny.

You may also begin to see some discharge and the shape of the doe's body may change as the babies begin to move into position for birth. Watch for behavior changes, such as pawing at the ground, loss of appetite, more talking, personality changes (e.g., the goat doesn't want you to leave), changing position frequently and looking uncomfortable, licking herself, breathing more heavily or grinding teeth.

Although some goats will isolate themselves, I have often observed fighting with other does, and actually had to separate the mother for safety. Each doe is different and may show different signs that she will kid soon.

According to David MacKenzie, in *Goat Husbandry*, as long as you can see the kid(s) as a bulge on the right side and see movement, the goat is unlikely to kid within the next 12 hours. I have not tried this method.

Kidding, or parturition, is normally divided into three stages:

- **First stage** of labor is when the uterine contractions dilate the cervix by forcing the placenta, fetus and amniotic fluid against it. This can last up to 12 hours in first-time moms, but is often faster for those who have previously kidded. Again, every doe is different.

- **Second stage** of labor is the period in which the doe pushes the kid(s) out. It usually lasts less than two hours, but can be longer.

- **Third stage** of labor is expulsion of the placenta and the reduction of the uterus back to its normal size. In most cases the placenta is passed within and hour or two after birth but in rare cases can take hours. The uterus does not reach its pre-pregnancy size until about four weeks later.

First Stage

Birth is controlled by a complex series of hormone releases. It starts with secretion of estrogen by the ovaries, which cause the uterus to contract. At this point, the kids will not be felt to move, the bulge in the doe's right side will change and the rump will begin to slope more. This may not be visible to any but the trained eye.

The doe will also become restless. If you have a clean kidding pen prepared, now is the time to move her there. Like all mammals, goats like a quiet, safe place to have their kids. It should be lit well enough (or have access to light) that you can see what you are doing if you need to help, but dim enough to be comfortable. It shouldn't be too small, so she can move around as the labor progresses.

Avoid putting water in the pen; kids have been known to drown in it and the doe will be focusing on the task at hand. If you do want to give her water, make sure it is warm and take away any that is left.

Sometime after this you may see a thick discharge. This means that the doe has lost her cervical plug. You will likely see a change in discharge as labor progresses. It thickens and changes color; in some cases it may be tinged with blood. This is normal. What is not normal is thick, rusty-brown discharge, which may indicate a dead fetus. If you have questions, contact your veterinarian or an experienced goat breeder.

Your doe, at this point, will probably be repositioning herself regularly, trying to get comfortable. She may start licking herself or objects, "mama-talking" (a special talk reserved for welcoming kids), or in the case of a very spoiled goat, she may demand that you stay there and pet her throughout.

Second Stage

The second stage of labor is where the real work begins. At this point, the babies have lined up for birth and the doe begins to push them out, in sync with the uterine contractions. The contractions become stronger and closer together. While I understand that some full-size goats deliver standing up, the minis and Nigerians that I raise prefer lying down. If they are tame enough to want me around, I will sometimes let them use my hand for a brace to push their back legs against. The doe may or may not cry out at this point.

The first sight that tells you the labor is progressing is what looks like a balloon at the opening of the vagina. This is the membrane surrounding the baby. The doe may start licking in earnest between pushes; sometimes situating herself so she can lick up the amniotic fluid. With some more pushes, you may be able to see two little hooves and a little nose, which indicates the baby is positioned properly. The kid is moving down the birth canal.

If you see just the nose and no legs, and the progress of the birth seems to have stopped, insert a thoroughly washed finger in to feel for bent back legs. You need to pull just one of these gently up to help make the birth easier, or in some cases, occur at all. This decreases the width of the shoulders. The kid should come out easily now with just another push or two.

I have found that Nigerian Dwarves are often born breech—back feet first—with no problems. Frank breech position, where the hind legs are folded underneath the kids, are a different story. This will need to be corrected prior to birth, which you can do by gently pulling the feet, and then the kid, out during a contraction. This prevents it from accidentally inhaling amniotic fluid and getting aspiration pneumonia or drowning.

Another presentation problem, which I have encountered only once out of hundreds of births, is crown presentation. This is where the kid's nose is pointing down toward the body, with the top of the head presenting. Because I didn't know what I was feeling, and the vet's hands were too big, we had to perform a c-section. (That kid was born four hours after his brother, and did just fine.)

Once a kid is born, wait for the umbilical cord, if it hasn't already broken, to stop pulsing. This will help to ensure a healthy kid, because of the blood that is still being transferred. (Obstetricians could learn something from this.) Once the cord breaks on its own, or collapses when the blood flow stops, you can tie it off securely with dental floss in two places: and inch or two from the kid's belly and an inch past that. Only now should you cut it.

All this time mom will be licking and cleaning the baby. If the doe does not want to get up or can't reach the kid, you can bring it to her. She will continue with this behavior until the next kid is ready to be born, which can be very quickly or take another hour or so. Longer times may be a sign of malposition, so if a placenta has not been delivered yet, and you aren't sure if there are other kids, you may want to check. Remember to err on the side of non interventionism, though. This is where experience comes in.

There are a couple of ways to check for more kids: First, you can check inside the doe with a finger. That will at least tell you whether another kid is in the birth canal and needing positioning help.

If that tells you nothing, you can "bump" the doe. That means standing being her and with your hands on the doe's abdomen lift up quickly to feel for another kid. A more effective, but also more invasive method is to check inside the uterus with a well-washed hand and forearm that has been lubricated. I have found that having a bucket of soapy water helps this effort immensely. Wash the perineum and be gentle with your exploration. A loose-feeling uterus will contain no other babies.

I have had to do this only once in my seven years of kidding experience. In that case, the doe had ringwomb, which means the cervix will not dilate enough, and I had to slip the cervical lip around the large kid's head.

Usually you will know that the doe is through kidding.

If you deliver a kid that isn't breathing and seems very weak, you can try blowing softly into the baby's mouth, or simply hold tight to the kid (one hand on the legs, and one on the neck to stabilize the head) and swing it back and forth in a 90 degree arc to clear the mucus. This is what I did with the c-section kid born four hours after his brother. If the kid is not able to suckle, you may need to tube feed it.

Third Stage

Once the kids are born, they should start nursing, which causes a release of oxytocin—also known as the bonding hormone. It not only helps mother and baby bond, but it stimulates uterine contractions that lead to delivery of the placenta and closing of the cervix. You will sometimes need to help the kids find the teats so they can nurse; in rare cases (once, in my experience) the mothers will not know to nurse their young. Breeders who pull the kids at birth should milk the goat, as this has a similar effect.

Some goats with multiple kids may have more than one placenta and, in rare cases, the first one may be expelled between deliveries of kids.

You should normally expect to see that the doe has a bag of amniotic fluid attached to some umbilical cord hanging from her vagina. The weight of the fluid apparently helps to pull out the placenta after it detaches from the uterine wall.

Failure to deliver the placenta may indicate that another kid is still in the doe. Never pull on the membranes to remove the placenta, as that can cause ripping and lead to problems later. You may obtain a prescription from a veterinarian for oxytocin for a retained placenta. Don't assume if you found already-born kids and didn't find a placenta, it hasn't come out. Goats, like mammals other than humans, normally eat their placentas. In researching the benefits of eating human placenta, I discovered a Russian study from 1954 in which women who were having trouble with milk production for breastfeeding were surreptitiously given placenta. Those who received the placenta improved their milk production, while those given placebo did not. Other evolutionary reasons for placenta-eating may exist, as well.

Once kids are born, dip their navels in 7% iodine to prevent navel ill. Make sure they are thoroughly dried, especially if the weather is inclement. They need to receive some colostrum within the first hour, if at all possible. Once mom has completed her job, I have a ritual of bringing hot oatmeal with molasses and a bucket of hot water to her. The water replenishes her system, and the oatmeal is a great treat with the added benefit of being galactogenic (helping to produce milk).

Make sure the kids have a cozy spot, with a heat lamp, if necessary, and then get the camera!

From *Ruminations* #52

A Difficult Kidding
by Cheryl K. Smith

I've been breeding and raising miniature dairy goats (Nigerian Dwarf and Oberian) since 1998. During this time I have been present for the births of more than 100 kids. Most of the time the goats do fine birthing by themselves; I have had only one c-section, which I attribute to not being experienced enough to diagnose and resolve the kid's crown presentation.

The first kidding this year was on February 11. I had determined that the doe would birth her kids within 24 hours, by feeling her tail ligaments. When they disappear into mush, you can usually guarantee that the birth will happen soon.

The doe, Mystic Acres Edinburgh, is a fairly large Oberian—a cross between a Nigerian Dwarf and an Oberhasli. Because the sire, Mystic Acres Alexander, is three-quarters Oberhasli and one-quarter Nigerian Dwarf, I thought perhaps Edinburgh was carrying two rather large kids.

Edinburgh is somewhat shy and clearly didn't want me in the kidding pen with her, so I stayed out during the day to let her progress. We had to muck out the barn that day, so I thought that perhaps she was holding back on labor until thing were quieter. We left the barn around 3:00 pm, coming back once for the evening feeding. Still nothing happening other than some bloody mucus. I turned out the lights to give her a better atmosphere for kidding.

As the day turned to evening, I listened to the baby monitor in the house.

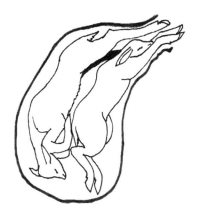

Around 9:30 pm I heard her making some small moans and decided to go back to the barn to check on her. She was very restless, constantly changing position. Around 10:10 I noticed that she seemed to be pushing, but still could not get comfortable. I told myself that I would wait until 11:00, not wanting to intervene too soon.

Around 10:30, as she was starting to get louder, I noticed some feet hanging out of her vulva. I went into the kidding pen to see what was up. Her pushing was clearly not progressing very well. At this point she seemed willing to get some help.

Upon examination, I found three legs coming out. Time for the bucket of hot soapy water and gloves. Diagnosing the problem requires feeling the kid(s) and visualizing what you are feeling. Was the kid sideways with front and back legs trying to come out? Or was it two kids trying to be born at the same time? The hind leg was a bit larger than the front legs, and I thought I felt two bodies. I decided it was two kids, pushed the hind leg back in and gently pulled on the two front legs. Out came a baby goat, head and legs first.

I knew that almost an hour had passed since Edinburgh had started pushing, but I had felt the legs moving when I did the initial examinations. But now the baby was limp and unresponsive. Mama began licking her and I rubbed her vigorously with a towel, trying to stimulate her into life. No response. I was sure she was dead. We continued the rubbing together and I made sure her nose and mouth were clear of mucus, but the baby didn't breathe or move. I picked up her little head and gave two small puffs into her mouth. Within a few seconds she took her first breath. Edinburgh and I continued to stimulate her and clean her up and she gradually came alive. What a relief.

After getting her cleaned up and wobbly on her feet, Edinburgh was ready to deliver the next kid. She stood there and easily pushed a long, breech kid out—feet first. He slid out easily and immediately began snorting and shaking his head. I moved him over onto a towel by the first baby and Mama went through the same routine with him. She spent another 20 minutes or so cleaning and stimulating these kids. I dipped their navels in iodine to prevent navel ill and made sure that they were dry enough to prevent a chill. Thinking that she was finished with the birthing and needing a break, I went back to the house to wait for the placenta to come.

I returned to the barn about 20 minutes later to ensure that the kids had figured out how to nurse and dispose of the placenta but found Edinburgh lying down again with hard contractions. She was trying to push a third baby out. I thought the drama was over and this would be an easy task. But 10 minutes or so later I discovered that she again had obstructed labor. I saw a little foot hanging out. That gradually became two feet, but no head. Again I washed my hands and got the soapy water, and put my hand in to check. I couldn't feel the little nose that should be coming out. Were these front legs or hind legs? They were clearly front legs.

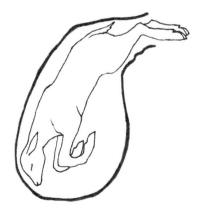

Again I felt inside: no head. I couldn't visualize where the head was, but finally decided that it was either bent to the side or backwards. I tried to push the legs in and feel the head, hoping to get it facing the right way. No luck and Edinburgh was in pain.

Driven by the knowledge that if I didn't get this baby out, both mama and kid would die, I pushed one leg in and gently pulled the other out, moving her perineum around it as the bent head gradually came out. It worked and she had only minor tearing. The baby was out! Alas, he appeared to be dead as well.

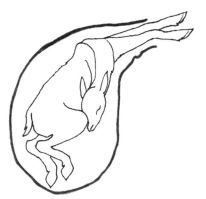

Again Mama and I began working together to stimulate the little guy into life. Breathe, baby, breathe! Suddenly he was trying to breathe, but was unable to get a breath. His tongue was hanging and he appeared to be gasping.

The next step was to swing him. This sounds rather harsh, but it is a technique I learned from the vet who did the c-section on my goat, Pearl. That kid was delivered four hours after his brother's normal birth and did just find.

I swung the little guy back and forth, holding onto his front and back legs and stabilizing his head. This gets the mucus out of the airway. When I checked him, he continued to gasp, little tongue hanging out. I tried this a second time, then took him back in to Edinburgh. Then I ran and got a syringe to clear his throat and nostrils. Finally the tongue went back in his mouth and he was breathing. We continued stimulating him by licking (Edinburgh) and rubbing with a towel (me).

The other two kids by this time were ready to eat, crying and wobbling around by mom. They managed to get started on her colostrum, while she tended to the third kid. He took a while to attempt to stand and was quite limp.

After waiting until 1:00 am for her to birth the placenta (which can take hours) and helping the third baby get his first colostrum, I decided that I needed to go to bed. Mama and kids were all on their feet. My last job was to get Edinburgh a bucket of hot water with molasses, fresh alfalfa and some hot oatmeal—which is lactogenic (milk-producing) and seems to hit the spot after the hard work of birth. She drank about a gallon of water, gobbled the oatmeal and nibbled at the alfalfa. She looked worn out and was probably in pain.

Goat Health Care

When I went out at 6:00 am the next morning I found two placentas on the ground and Edinburgh and her kids curled up together. I discovered that we had two brown-eyed boys and a blue-eyed girl, all healthy and strong. What an introduction to the 2007 kidding season! Only 12 more to go....

From *Ruminations* #57

Other Kid Positions

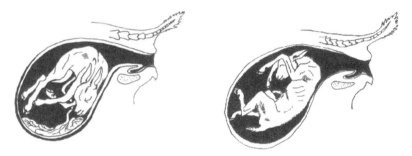

Two more positions that kids may be in. Above left is a transverse lie; to the right is a shoulder presentation, with both head and feet down. Both are unusual.

The method illustrated to the right is sometimes used to pull a kid's nose up when it is presenting crown (top of the head) first.

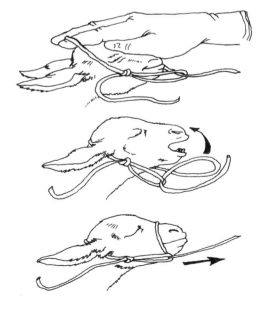

A Difficult Kidding: Part II

In 2008, my goat Mystic Acres Edinburgh was bred again and, like the prior year, she was huge. Her labor began innocuously enough, but sometime into it, I noticed that she was acting uncomfortable. She would lie down, push a little, then get up and walk around. It looked like something was going on, so I decided I would check her. I washed up, got my bucket of soapy water and entered the kidding pen.

The first thing I saw at this point was a first for me: A little tail was hanging out! I knew what I had to do—this was a frank breech and I had to get the kid's legs pulled back as gently as I could. This required pushing the kid back in a little and then gently moving the legs so that they were the first part out. I wanted to be careful not to damage her uterus in any way.

The kid seemed in no distress, and I helped mom get her cleaned up. Then Edinburgh was down again, pushing. I saw one back leg, inside an amniotic sac, but something seemed to be holding it up. The amniotic sac was ballooning out, but the kid was making no progress. I thought I would need to get the second leg down. I reached in again to check her and was uncertain what I was feeling. I had to focus and really think about it. I realized that I was feeling the head of another goat whose legs were back. Now what? Two coming out at once. Which should come first?

Being careful not to break the amniotic sac (it could cause distress in or even kill the kid that I decided needed to be born third), I pushed the leg back into Edinburgh. The balloon of amniotic sac still hung out. I pushed the head in a little, and then gently bent the front legs forward at the knee and the baby came out effortlessly.

Then, with little time between, I checked the third goat and found that one of the back legs was not coming out. I got that leg down and then, with quite a bit of effort, Edinburgh pushed her out. This goat was probably twice the size of the first two, and a female.

After another round of cleanup and a short rest, Edinburgh again laid down and began to push. Out came a fourth baby—head first and feet down—a bonus doeling!

When it was all over we had three beautiful doelings and a buckling, a contented mom and a goat keeper with a lot more experience in goat midwifery.

Acquired Birth Defects in Kids
by Michele Konnersman, DVM

Hormonal causes

Freemartins. Doe kids that have a male twin can be born as freemartins (a type of hermaphrodite). This occurs relatively infrequently, but is caused when the blood supply is shared between the doe kid and her male sibling, and testosterone flows into the doeling, inhibiting development of the reproductive tract. Doe kids born with one or more male siblings can be checked for the freemartin condition by measuring the length of the vagina. Freemartins usually have a very short vagina when compared to a normal doe kid of similar size.

Toxic causes

Albendazole, a wormer also called Valbazen, is toxic to the fetus when given to pregnant does, especially on Day 12. Kids can be born that have parts of the skull and brain missing.

Another wormer, **oxfendazole**, also called Synanthic, can cause hare-lip, uterine aplasia in doe kids, fused ribs, and eyelid defects, especially when given on Day 17. A doe kid with uterine aplasia will never be able to conceive.

Skunk Cabbage Poisoning, by *Veratrum californicum,* can cause cyclopia in kids if the pregnant doe eats the plant on Days 14 and 15. The toxin causes deformities of the fetal skull such as one central eye, or a monster fetus with defective nose and the eyes close together.

Locoweed, *Astragalus sp.,* causes small, weak kids. The legs of affected kids are deformed, with carpal flexion being the most common sign, along with rotation and contraction of joints.

Lathyrism is caused by the **Lathyrus species of peas**, also called vetchling, wild pea and flat pea. Scientific names include: *Lathyrus ochroleucus L., L. japonicus, L. palustris L., L. venosus Muhl, L. aphaca L., L. tuberosa* and *L. latifolius L.* The poisonous parts of the plant are the seeds, but the foliage can also contain the toxin. The vetchlings flower in the summer months, usually from June through September. Does that eat the seeds and plant in the first three weeks of pregnancy can have kids born with contracted or over-extended pasterns and carpal joints, rotational deformities of the legs, crooked spine, hump back or wry neck.

Mineral deficiencies

Copper deficiency can cause kids to be born completely paralyzed, or kids may walk with a strange gait. The condition is sometimes called "swayback" or "swingback" due to the way the kids have to move to walk, by swinging their rear from side to side in order to take steps.

Iodine deficiency in the pregnant doe can lead to kids being born with very short or fuzzy hair and very large goiters. The thyroid gland may be so large that the upper neck is larger than the head. This can cause birthing difficulties. Pregnant goats must have a source of iodine at all times, usually in a free choice mineral mix provided by the keeper.

Certain plants are goitrogenic, i.e., they will cause a doe that eats too much of the plant to be iodine deficient. Plants that are goitrogenic are the plants of the **Brassica** family, including mustards, wild radish, broccoli, kale, turnips, cabbage, and also soybeans. The goitrogenic substance in soybeans is only partially destroyed by processing soy into meal.

Selenium deficiency can cause heart defects and stiff and painful rear leg muscles in newborn kids. Affected kids can also have poor sucking and swallowing abilities. Sudden death can occur from the cardiac form, but usually a kid will have a rapid heart rate, rapid respiratory rate, increased respiratory effort and weakness which can appear as if the kid has pneumonia.

Infections

Toxoplasmosis infections in kids before birth can cause blindness in newborn kids. Toxoplasmosis is spread by barn cats, particularly kittens, so spaying and neutering is very important.

Bluetongue virus can affect goats, although the infection is uncommon. It can cause kids to be born with abnormal brain development, water on the brain, poor balance and shortened bones of the forelegs.

Border Disease, also called "Hairy Shaker Disease," is caused by a pestivirus of the Flaviviridae order. It can cause kids to have poor balance, tremors and retardation. They are usually born undersized and have an excessively hairy birth coat. Their skulls can show some deformity and they can have very short lower jaws.

From "Ask the Vet," *Ruminations* #51

Kid Care
Post Kidding

Protect the Navel

If the umbilical cord is not broken or is long, cut it with sterile scissors or knife at least an inch from the belly. If it is bleeding, tie it off with dental floss close to the navel. Dip the entire length of cord in 7% iodine.

> **Trick:** Some vets recommend that each kid be given orally the contents of a 1000 IU vitamin E gel cap as soon as possible after birth. Often the kid has some selenium available but not enough vitamin E to effectively use the selenium, as they work together.

Bo-Se Injections

If a kid seems weak after birth, try giving it ½ ml of Bo-Se SQ. In most cases the kid will perk right up. Some breeders routinely give this to their kids.

Provide Colostrum

Colostrum contains important nutrients and antibodies. Try to ensure that the kid gets colostrum within the first hour after birth. In some cases, this may mean milking out some colostrum from a full udder and helping the kid nurse.

> **Heat-treating Colostrum:** Put in small zip lock bags or jars, bring to 135° F. and hold for one hour. Do not let the temperature go above 140° F. or below 130° F. You can use a crockpot or Weck canner with a water bath.

If the kid is to be bottle-fed, warm the heat-treated colostrum to body temperature. Those who bottle-feed should keep heat-treated, frozen colostrum from the prior season or from one of the first does to kid, to feed the first kids.

If the kid is unable to nurse; try a bottle. If the kid is unable to suck at all for any reason, tube feed some colostrum.

CAEV Prevention

In cases where does are known to have caprine arthritis encephalitis virus (CAEV), kids need to be taken from the mother before they are licked and before they contact each other. Because the birth fluids contain CAEV infection, each kid must be washed in mild soap and warm water, then dried thoroughly.

I also recommend that each kid be kept separated and tested several times, to weed out any kids that have been infected during birth. They should be kept separated because they can transmit to each other in the pen. If you suddenly find that one is infected, then all kids in that pen may have been infected from sharing nipples on lamb bars and feed pans.

Finding a source of cow colostrum from older cows that are Johnes-free may be easier than heat-treating goat colostrum. Cow colostrum may be frozen months ahead so that it is readily available if needed.

Trick: Freeze colostrum from CAEV-negative does, or heat-treated colostrum from CAEV-positive does, for orphans or kids born the next year.

Otherwise, feed all kids goat colostrum and milk. I believe that by paying close attention to detail, raising kids that are CAEV-free is still possible.

From "Ask the Vet," *Ruminations* #50

Housing

For goat breeders who are bottle-raising their kids,

> "[L]arge cardboard boxes with clean bedding material work well for housing newborn dairy goat kids, especially in large herds. A doe's kids can be placed in one box, and the dam's identification written on the box as a means of identifying kids until they can be labeled with paper collars and permanently identified by tattoo. Disposable boxes are a useful means of preventing buildup and spread of enteric pathogens. Kids can be kept in these boxes for about two weeks, after which the box can be destroyed and kids housed in larger groups."
>
> Veterinarian Joan Dean Rowe, North American Veterinary Conference, January 2006

Tube Feeding Tips

- Place the kid sitting up, not on his or her side. If necessary, fold up a towel to prop up the kid's head. This prevents aspiration, in the event the kid coughs up any fluids.
- Clean and sterilize supplies after each tube feeding.
- You may have to tube feed a weak kid just one time to get it up and energized. If more tube feeding is required, it should be done only every 2–4 hours with the same small amount. Frequent, small feedings are better than infrequent large feedings.

Tube Feeding a Weak Kid
by Cheryl K. Smith

If you are breeding goats, sooner or later you will find it necessary to tube feed a kid. When a kid is born too weak, is unable to suck, or is too sick to nurse or drink out of a bottle, it becomes necessary to get fluids into him or her another way in order to ensure survival.

Tube feeding is preferable to feeding with a syringe. Syringe feeding can sometimes cause lethal results and trauma to the person doing the feeding. This is because feeding with a syringe can easily lead to aspiration—that is, breathing the fluid into the lungs. The kid may inhale the fluid, which will cause aspiration pneumonia or choking to death.

A variety of diseases and conditions may prevent a kid from nursing. Malpresentation at birth, problems with labor leading to oxygen deprivation, or prematurity are among these. If you are not present at birth, especially with multiple births, the doe may not pay attention to the kid, allowing it to get chilled. Other conditions that may lead to the need for tube feeding are enterotoxemia, coccidiosis, polioencephalomalacia and *E. coli*.

What do you need to tube feed a kid?

If the kid is a newborn, the ideal thing to use is colostrum that you have frozen just for times like this. To get a weak kid going, mixing some molasses or Nutridrench with the colostrum is helpful.

If you don't have frozen colostrum or another doe with fresh colostrum, the next best thing is electrolytes, B vitamins, probiotics, goat milk or milk replacer. I keep on hand not only frozen colostrum, but dried electrolytes I got from my vet. If the goat is being tube fed because of scours, electrolytes are the best bet. Stay away from milk or milk replacer for at least a day.

You will also need a feeding tube and, ideally, a 60 ml syringe with an irrigation tip. Have a bowl of clean warm water handy and a small syringe. I also like to have two people, as it is a lot easier both emotionally and physically, particularly if you have a large kid that wants to fight.

How do you get the tube in the stomach?

The first step in tube feeding is measuring how far you will need to insert the tube so it ends up in the kid's stomach. Measure from the nose to the center of the ear. Then measure from the ear down to the chest floor. Add the two measurements and mark the tube at that point.

Have someone hold the kid securely and dip the end of the tube in the warm water so that it is softened. Insert the tube into the kid's mouth, over the tongue and down the throat until you are at the length you marked. You may be able to feel the tube as it is passing down the esophagus. Very weak kids will not even struggle; bigger ones can fight you.

If the kid was crying before you inserted the tube and suddenly stops during the process, pull the tube out until it can cry. Then try again. Tube feeding into the lungs can cause pneumonia in a kid; more likely it will die.

Several methods are suggested for determining whether the tube is inserted correctly. The first is to smell the end of the tube for the milk smell of the stomach. This obviously will not be true for newborns, who have not had any milk. Another method is to listen at the end of the tube for little crackles. If you hear breath sounds, you will need to withdraw the tube and start over. The third way is to place the end of the tube into a cup of water. If it blows bubbles, you are in the lungs and will need to try again.

What is the process for tube feeding?

Put 2–4 oz of the liquid you will be feeding into the tube (I use 2 oz for Nigerians). Attach the syringe to the end of the tube. Make sure you have completed the checks above to ensure that the tube is actually in the stomach. Put about 5 ml of water in the syringe, checking to see that it flows down the tube. If not, withdraw it a few inches and replace it. This is to make sure it is not against the stomach wall or twisted in some way.

If the water flows down the tube, add the feeding liquid to the syringe. There is no need to use the inside part of the syringe; gravity will deliver the fluid to the stomach if you hold the syringe above the kid.

After administering the fluids, add another 10 ml of water to rinse the syringe. This step is not always critical, but it does help to prevent aspiration of milk or electrolytes, when the tube is being removed.

How do you remove the tube?

First remove the syringe from the end of the tube. Then remove the tube SLOWLY while covering the end to prevent any excess fluids from getting into the lungs. Removing the tube too fast can cause discomfort to the kid, as well as potential tissue damage.

From *Ruminations* #33

Floppy Kid Syndrome

Floppy kid syndrome (FKS) is a sudden onset, in an otherwise healthy young kid, of extreme weakness and inability to move the legs. It is associated with metabolic acidosis (increase in overall acid in the body) in which no specific organ systems are abnormally involved. No gastrointestinal or respiratory clinical signs, such as diarrhea, dehydration, or difficulty breathing, are seen. It usually occurs late in the kidding season.

FKS was first documented in 1987, but the cause is unknown and further research is needed. There is no difference in incidence between dam-raised and bottle-raised kids, or those given pasteurized or unpasteurized milk. It can spread rapidly among young kids. Unfortunately, in some cases, a number of kids die before the goat breeder determines what is going on.

Gastrointestinal disease is strongly suspected to be the cause for this syndrome. Early untreated cases, where death occurred, should be necropsied to help determine the cause.

Affected kids are normal at birth and then develop sudden profound muscle weakness at 3–10 days of age. Kids are often reluctant to nurse, but can swallow. Biochemical findings include metabolic acidosis, decrease in bicarbonate, normal to increased chloride and occasional hypokalemia (low potassium). No gastrointestinal or respiratory dysfunction is apparent (e.g., diarrhea, dehydration, or breathing difficulties).

Tip: Treat kids with suspected FKS by giving ½ teaspoon baking soda in cold water.

Diagnosis is made based on supportive clinical and laboratory findings. Any disease that can lead to a profoundly weak or acidotic kid, such as white muscle disease or enterotoxemia can be mistaken for this syndrome.

The best treatment depends on early detection of the problem. Less severe cases can be treated by giving the kid baking soda as soon as possible. The recommended dose is ½ teaspoon in cold water given orally. If a kid is unable to nurse or drink a bottle, he or she may have to be fed by stomach tube. You should see improvement within two hours, if the kid actually has FKS. In severe cases, intravenous fluid and bicarbonate administration will be required.

According to the literature, there is no association between treatment with antibiotics or vitamin/mineral supplements and recovery. In fact, some kids may recover with no treatment at all. In other cases kids may relapse or take up to a month to recover their neuromuscular function.

Excerpted from *Ruminations* #44

Correcting Hyperflexed Legs in a Newborn Kid

by Margo Piver

My friend Kelly and I were at a Nubian breeder friend's house having bucks collected. While we were there we went in their kid barn to help them feed. They had a little buckling that was having a horrible time trying to stand. His hind legs were knuckled over at the fetlock joint and the hocks hyperflexed forward like Barbie doll knees. Kelly is a sucker for babies and especially sad little ones that need some TLC. They told her she could have the buckling if she wanted him, so of course he came home with us.

On the drive home I was thinking about how we could splint the legs to keep the hocks from flexing back in such a painful way. We thought about ways to brace the legs with toilet paper rolls and such. These would work great but then the leg would not bend for the kid as he slept.

Finally it dawned on me that all we needed to do was to tape a strap to the front of the leg at the thigh and below the hock on the cannon bone. This way he could fold his leg up to sleep yet the strap would tighten up before the leg was able to straighten out too far, forming an artificial tendon on the front of the hock.

Here's how we splinted his legs:

We folded a piece of duct tape in half to form the tendon.

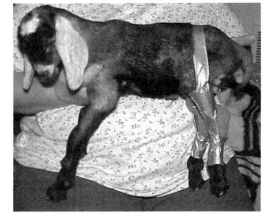

We wrapped the leg in vet wrap and taped the tendon to the front of the hock above and below with the tendon tight when the kid stood in the correct position.

To keep the tension from pulling the tendon out of position, we made suspenders out of another folded piece of duct tape. The suspenders were taped to the top of the tendon on both sides.

After applying this to the kid he was able to stand and his legs stayed in the correct position for the next few days while they got stronger.

From *Ruminations* #38

Navel Ill

by Jennifer White

Navel ill is a disease caused by infection of the umbilical cord. It is usually caused by an unclean environment. Right after kidding, a kid's fleshy navel is susceptible to infection. The navel contains blood vessels which, if infected, can create an infected liver or possibly blood poisoning. If the infection spreads to the joints, it causes joint ill.

Trick: Pour iodine in an old prescription medicine bottle for dipping kids' navels right after birth.

Some symptoms are a swollen, painful navel. It may look red or angry. Commonly the kid is off his/her feed and may have swollen joints. The swollen joints usually become evident after a few weeks.

Generally navel ill is treatable. Your vet will probably prescribe antibiotic injections. Clean the area around the navel with antiseptic iodine and remove rusty scabs by soaking with a hot, moist cloth. Drain pockets of pus, if there are any.

Your best defense against this problem is to provide a clean kidding area and dip the navel in iodine, or spray with anti-bacterial spray right after delivery.

From "Kids and Kids," *Ruminations* #34

Feeding Kids

The ideal for kids, as for any mammal, is mother's milk. If that cannot be done (e.g., the dam is ill, dies or rejects the kid), pasteurized goat milk is second best. Many goat keepers have good luck with milk replacer and others use cow milk, often with Half & Half added for butterfat.

The amount fed to the kid will depend on whether it is a miniature or standard breed and its age. Kids of standard breeds usually drink from 12–24 oz a day for the first week and then up to 16 oz a day for the next nine weeks, until weaned. Feed 3–4 times a day initially (if time allows) and then decrease gradually to once or twice a day before weaning. Some breeders wean kids at eight weeks and others wait as long as 12 weeks.

After the first couple of weeks, the kids will begin to mimic others in the herd, eating hay and drinking water. In some cases where the kids are dam-raised, they may not wean themselves for months. This can be a problem in that the kids may get too fat, if they have an indulgent mother. This can lead to milk goiter and problems with pregnancy when they are bred.

Why Disbud?

by Roxanna Gill

Editor's Note: People often argue that horns on goats are "natural" and should therefore be kept. The following is a list of reasons to disbud. Remember that in order to show your mini goats they also must be disbudded.

Having some goats with horns in my herd, I wanted to share my experiences. No one likes disbudding, however, if done properly it causes a lot less pain than a goat with horns may have to suffer.

1. **They get stuck in fences.** Mine get left alone by the other goats when they get stuck, but I could easily see that an aggressive goat in the herd could beat on them until they are dead or wounded badly. Or they could be killed by a predator. They never pick the shady areas to get stuck and they always seem to get stuck in the middle of the night or during the hottest part of the day. They keep going to the same area and getting stuck over and over, too.

2. **They will accidentally hurt you.** We don't keep aggressive goats in our herd but no matter how careful you are, at some point you will get accidentally hurt. I've come close to being hit in the eye; have gotten caught up under the ribs by horns, when a goat misstepped and caught me while falling; have gotten butted when the goat was aiming for another goat; and I have had numerous scrapes, etc. Trust me, it's no fun. The first time is not such a big deal, but being punctured or getting caught in the wrong place can be very painful.

3. **They seem to be more aggressive** to the other goats and hit a lot harder than the disbudded goats.

4. **They tear up fences, housing, etc.,** scratching between the horns and "sharpening" them. We do have one buck that sharpens all the time. Some use their horns like a pry bar to tear stuff apart.

5. **The horns can break.** We have had several break off part of a horn or the entire thing. If you don't like disbudding, you definitely will hate and be horror-stricken when you have to cauterize a broken horn. If a horn breaks when no one is around, the goat can bleed to death.

6. **You have potential financial liability** if another person is injured. I don't want to be responsible for not protecting the safety of others.

7. **The sales potential is much better** if they are disbudded. 90–95% of people won't even look at a horned goat and at some point you will want to/have to sell some of the babies or even a goat that excels otherwise.

From "Just Browsing," *Ruminations* #51

Disbudding— A Learn-as-you-go Adventure

by Cheryle Moore-Smith

We decided to start disbudding our goats about 13 years ago, which was also when we moved to Cape Neddick, Maine. Being new to the area we had not established a working relationship with a local veterinarian yet.

We started asking around about which veterinarian would be an appropriate one for our little goats when a "helpful" guy we met at a feed store told us he had goats and disbudded kids for other folks all the time.

That next weekend we took two of our kids to his house. All that happened was that the "helpful" feed store guy managed to burn a little flesh, and scare the you know what out of our kids *and* us. It was evident that he did not know what he was doing. The kids started to grow horns.

After that experience we made an appointment with a vet who came highly recommended. He told us that the horns that were coming in on those two kids were too big to disbud. So, we found homes for those two kids with folks that wanted horned goats. We also decided that veterinarian was too far away to be a practical choice, especially if we ever needed him in an emergency situation.

By the time the next kids arrived we had located a good veterinarian closer to us. He preferred to anesthetize the kids and cut out the horn buds. He did it right in our kitchen and gave the kids a reversal to the anesthesia when he was done. We had our vet come several times and use this procedure, although we were frequently frustrated by does insisting that the kid wearing the vet wrap cap and smelling of blood and alcohol couldn't possibly be hers (this was before we started bottle-feeding a lot of our kids). We had very few problems with scur growth using this method but it was too expensive to have the vet out each time. It became evident that we would "have to take the goat by the horns" and do it ourselves. We did not want to deal with the blood though so knew it was time to try a different method of disbudding.

We purchased a Portasol cordless butane powered disbudding iron. We also purchased the optional "pygmy tip" but found it to be too small and therefore inadequate.

We knew that we wanted to continue anesthetizing the kids so the vet agreed to provide us the xylazine (also referred to as Rompun).

At this point I wish to make note that when using anesthesia it is extremely important that you do not overdose. I know that several folks would strongly disagree with our disbudding method; BUT IT WORKS! Each person has different methods of doing things and it is important to remember that what works best for one person, may not work best for another. It is also *even more* important that you have a strong understanding of how to properly anesthetize goats before you begin using this method of disbudding. I would suggest that before you attempt to do this procedure on your own, that as we did, you work closely, and several times, with an experienced vet or another experienced breeder who is familiar and comfortable with this method. Although there are risks involved whenever anesthesia is used on any animal, we have at this point disbudded hundreds of kids using this method and have never lost any or even had the slightest bit of side effects with any of them.

Tip: Disbud full-size and miniature goats earlier than Nigerian Dwarves and Pygmies. Bucklings can be disbudded as early as two days old.

We are constantly tweaking the procedure but here is what seems to work best for us:

- Disbud (Nigerian) does between the age of 5–10 days and bucks between 4–8 days. (I use age as a general reference here but it is actually according to when the horn buds feel the appropriate size for disbudding, which you will get more familiar with as you attain more experience.)

- Light your disbudding iron and put in safe place allowing it to reach the appropriate temperature while doing the first few steps below.

- Give the kid 0.001 or 0.002 ml of xylazine. (Better to underdose. Your goal is not total sedation, but so that the kid appears to be sleeping yet stirs when moved.)

- Shave the kid's head (we use a Golden A5 Clipper with a surgical blade).

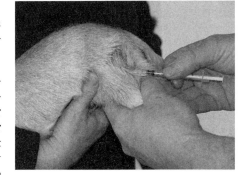

- Administer shot of tetanus antitoxin unless the dam was vaccinated during the last part of her pregnancy. We vaccinate our pregnant does usually about two weeks before their delivery due date but still vaccinate the kids with the tetanus antitoxin anyway. It can't hurt.

- Place iron over horn bud, slightly towards the center and front of head.

- Press down firmly while twisting the iron (This should be a very deliberate motion. Times will vary but it should take three or four seconds; the key is to be very deliberate and not hesitant with this motion so that you do not need to keep repeating it. Repeated applications no matter how quick they may be are the easiest way to build up heat in the skull and cause brain damage.

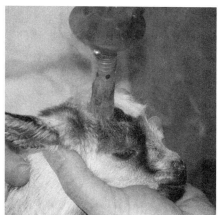

- The next step is one I consider very important, but unfortunately, many people skip it. It is to peel the horn bud cap off; it should easily come off with little or no blood. Sometimes for easy removal, you need to reapply the iron to one side of the visible ring made by the hot iron. You can use a pair of needlenose pliers to remove horn bud cap. We use the pliers part of a "Leatherman" tool (make sure to dip them in alcohol before using).

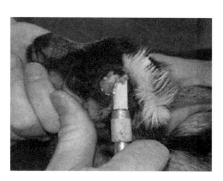

- Quickly apply the side of hot iron to the exposed area to cauterize.

- If the horn bud is quite large once you peel the horn bud cap off (this will be a large bony bump under the cap that you have just removed), you can use a sterilized scalpel (an Exacto knife works well), and scrape the bud off of the skull. The bud is actually quite soft and will scrape off easily AND *this is probably the most important part of the procedure, and will ensure that you don't see scur growth later.*

- In some cases, especially if you have waited longer than you should have to tackle the disbudding or if you have a growthy buckling, you will want to make a second burn in front of the first and a little more towards the center of the skull so that you are creating a figure eight. Then as in the directions above, remove the second horn cap and follow same procedure of scraping and cauterizing.

- You may once again need to quickly apply side of hot iron to the exposed area in order to cauterize.

- As soon as possible, apply Blu-Kote to the exposed area. We use the Blu-Kote that comes with a dauber attached to the top of the bottle and *not* the spray. With the spray you run a risk of getting it in their eyes. I always use one hand to cover their eyes while applying.

- Set the kids aside in quiet spot where they won't get chilled and let them sleep. Some will already be starting to wake by the time that the procedure is done; others may sleep for 10–45 minutes. Keep checking to make sure that they are breathing okay. There is a reversal drug to the anesthesia. We choose to not use it as we believe that allowing the kids to sleep for a while lets them get through the worst part of the pain, and in doing so they seem to have a shorter recovery time. Also the reversal drug is the most expensive part of the procedure.

- Your kids will want a bottle as soon as they awake (if they are bottle-fed) but we hold off until they are fully awake so that they won't inhale any milk. (Editor's note: You can put them on mom to nurse once they are fully awake, if they are dam-raised.)

- We have gotten in the habit of putting a half of an aspirin in their bottle to prevent any possible swelling.

- In the last couple of years we have started to tattoo our kids at the same time as disbudding. Just as with the disbudding, the job is made a whole lot easier by dealing with a snoozing kid instead of an unhappy, thrashing kid.

You're done! I know that in being detailed, I've made it sound quite complicated, but like anything, once you get the hang of it you will realize just how easy it is. We have become quite proficient at this method of disbudding. We also have taught a few other breeders how to use this method and now they are disbudding their own kids the same way.

From *Ruminations* #38

Trick: Goats, particularly males, often develop scurs if they are not disbudded early enough or well enough. Kate E. Cole, of Elmira, Oregon, recommends cutting a piece from a child's "water noodle" and attaching it to the scur with duct tape or glue to keep it from growing into the goat's head.

Castrating Your Buckling

by Cheryl K. Smith

If you're breeding goats, it won't take long until you reach the point that you have to decide whether (and how) to castrate a buck kid. The following will help you make those decisions.

Should I castrate my buckling?

When deciding whether to castrate a buckling, there are a number of things to consider. First, not every buck is of good enough quality to be retained. To determine whether the buck should not be wethered, ask yourself whether his dam measures up. Is she a proven milker? What does her udder look like? Has she held up well over the years (if she is old enough)? Is her overall conformation good? What did the buckling's granddams look like and how did they perform in the show ring and milking?

The second thing to consider has to do with space. If you have more than one buck, you will probably decide that they should live in an area separate from the does and kids. This is a decision that will probably need to be made early on, since some kids (Nigerians in particular) have been known to breed as early as two months old. So do you have the space?

Should you decide to keep them, you may need another separate housing space for the young bucks, as well. While it may be tempting to put them in with the older bucks, be careful, as it could cause injury or even death. I had one buck with a broken back leg caused by another more aggressive buck, although they were housed together only when both were adults.

The third consideration has to do with smell. While you may not have difficulty finding someone to take a buck that you had later decided you didn't want, some people have no idea how bad they may smell. This is less of a problem with Nigerians—I find that they only smell really bad when there is more than one. In addition to the smell, bucks have other unsavory habits that could eventually lead to a less than happy home.

Finally, wethers are the most charming and sweet of all goats. They can stay with their mothers, they are very friendly, and they lack the aggressive drive of bucks. In many areas, the pet market for miniature goat wethers is still growing. So, although they bring much less money than bucks, at least you will be able to sell them.

When should I castrate?

Everyone seems to have an idea about when it is best to castrate a buck. While early castration is purportedly less traumatic to the goat, it also makes him more susceptible to urinary calculi (urinary stones). This is because early castration prevents the urethra from developing to full size.

I normally castrate between eight and 12 weeks—after "buckiness" begins to be a problem. I once waited for four months, and found it to be the least traumatic banding I had ever witnessed. The goat showed no signs of discomfort whatsoever. (On the other hand, for larger kids, testicle size can be a problem if too much time goes by.) Given a choice, I believe that kids prefer to stay with moms than to be separated and castrated later.

How do I castrate my buckling?

If you only have a few goats and/or are squeamish, you may want to take him to the vet to have it done. Whichever method you choose, give the kid a shot of tetanus antitoxin first.

I also choose to give my kids an aspirin and some valerian tincture, mixed with molasses, prior to the procedure.

There are three main methods of castration: surgical, emasculator and elastrator.

Surgical Method. The surgical method (cutting with a knife or scalpel) is considered the most reliable and inexpensive method. It requires a sharp knife or scalpel, soap and water, disinfectant such as Betadine, a syringe and 1 ml tetanus antitoxin.

Before starting, wash your hands and the instruments with soap and water; then disinfect or sterilize with boiling water.

Have another person hold the kid by hind and front leg with each hand, facing the person who will perform the castration. Wash the scrotum and disinfect it.

Push the testes up and out of the way and cut off the lower third of the scrotum with a cut parallel to the ground. Once the testes are visible, grasp one and pull it downward, holding firmly. Use the knife or scalpel to cut off the cord. Do this by scraping, rather than cutting cleanly. Make sure that you don't let the testis or cord re-enter the scrotum, to avoid infection.

Repeat this process with the second testis. Remove any cord that hangs out of the scrotum, to prevent entry of bacteria. Pour the disinfectant on the castration site.

The main disadvantage to this method is that it leaves an open wound.

Emasculator Method. For this method you will need an emasculator, also known as an emasculatome or Burdizzo, which is an instrument that crushes the spermatic cord. This destroys the blood supply to the scrotum, which leads to atrophy (shriveling) of the testes. It is bloodless and doesn't even break the skin.

The disadvantage of this method is that sometimes the cords are not completely crushed.

Hold the kid as you would for a surgical castration. Wash the area and disinfect it. Grasp the scrotum in one hand and push the testes down into the scrotum and the spermatic cord between your fingers.

Crush each spermatic cord (one on each side) separately. NEVER attempt to do both at the same time.

Place the emasculator on the upper scrotum, just below the teats. Position the jaws so that about 2/3 of the scrotum is crushed when the jaws close. Squeeze for about 20 seconds. You should hear a crunch. Do this twice, then repeat the procedure on the other side.

This method may break the skin, so check the kid after the procedure and apply antiseptic, if needed.

Elastrator Method. This is known as "banding" and is probably the most common method for castrating goats. Some people consider this method inhumane, but my observation has been that the kid feels discomfort for usually no more than two hours after the procedure, and some show no discomfort at all.

> **Trick:** Keep elastrator bands in the refrigerator to keep them from breaking down.

Once a band is applied, the scrotum and testes will fall off in 10–14 days. I examine the kid regularly for infection, and have sometimes had to cut the band and withered testes off, especially with older kids.

Banding requires a device known as an elastrator and castrating bands or rings. These bands are quite heavy and should be kept refrigerated to prevent them from breaking down.

Because this method is bloodless, washing and disinfection is not necessary. Have an assistant hold the kid in her lap, facing out. Place the castrating band on the prongs of the elastrator and turn it with the prongs facing the kid. Expand the ring by squeezing the elastrator.

Place over the scrotum and testes, positioning it close to the body, taking care not to squeeze the teats.

Push down the testes so they are below the ring. Release the elastrator and remove the band so that it is properly positioned. If both testes are not below the ring, or it seems not to have gone on properly, cut the band and start over. I once had to do this three times before getting it right.

Conclusion

All of these methods will be stressful for the kid, so if you are dam-raising, take him right to his mother to nurse, and if bottle-feeding, have a bottle ready to go.

Keep the kid out of mud and rain and give him plenty of room to run around, once he recovers from the initial shock.

And remember, everyone will like him better as a wether!

From *Ruminations* #39

Local Anesthesia for Castration by Elastrator

Wool and Wattles, the newsletter of the American Association of Small Ruminant Practitioners (AASRP), reported on the use of local lidocaine for rubber ring (elastrator) castration in young lambs. The pain levels in the animals was assessed through observation of behavior and measurement of their cortisol. The authors of the study, which was reported in *Veterinary Journal*, noted that local lidocaine was not as effective for pain relief in burdizzo castration.

The researchers injected lidocaine diluted with saline into the spermatic cords and under the skin around the scrotal neck of the lambs. They then waited five minutes before doing the procedure.

The authors suggested that waiting 10 minutes, rather than five, might have improved the anesthesia. They also noted that animals without the lidocaine showed much active abnormal behavior for about two hours after castration. The burdizzo-castrated goats showed more pain at the time of castration, while those with the elastrator method showed more discomfort after castration.

This may be a method that breeders can work with their vets to implement in goat wethers-to-be, to minimize their pain and discomfort with castration.

Identification for Goats: Tattooing and Microchipping

by Cheryl K. Smith

Tattooing or microchipping? Now that microchipping has come into popular use, more breeders are turning to this method for permanent goat identification.

Pros and Cons

Microchipping is accepted by the Nigerian Dwarf Goat Association (NDGA) and American Goat Society (AGS), but is only approved as a supplemental method by the American Dairy Goat Association (ADGA). The two mini dairy goat associations also both allow it. It's less messy than tattooing. It's also permanent, and you can't put the number in backwards.

However, you need a reader to determine whether the permanent identification number matches that on papers. The initial equipment is also more expensive than that needed for tattooing, costing around $400. As your herd grows, you will also need to purchase more microchips. In rare cases, microchips have been known to cause tumors in animals.

Tattooing is the traditional method. Supplies and equipment will cost about $50 to get started and only a few dollars a year after that. Tattooing also makes easier identification of a stolen or lost goat because it is visible to the eye. Tattooing takes longer and causes more pain to the animal than microchipping, however.

> ### Microchips and Tumors
>
> Various reports have linked microchips to tumor development in animals, including dogs, rats and fruit bats. A study of rats with implanted microchip identification devices showed that approximately 1% developed cancerous tumors in the area of the implant.
>
> Vascellari, M., E. Melchiotti and F. Mutinelli. 2006. Fibrosarcoma with Typical Features of Postinjection Sarcoma at Site of Microchip Implant in a Dog: Histologic and Immunohistochemical Study. *Vet Pathol* 43:545–48.
>
> Elcock, L.E., et al. 2001. Tumors in long-term rat studies associated with microchip animal identification devices. *Exper Toxicol Pathol* 52(6): 483–91.
>
> Siegal-Willott, J., et al. Microchip-Associated Leiomyosarcoma in an Egyptian Fruit Bat. 2007. *J Zoo Wildlife Med* 38(2): 352–56.

Microchipping

To microchip a goat you will need the microchip in its individual injector, a scanner and a form for recording the goat and microchip number.

Injecting a microchip requires three scans with the microchip scanner to verify that you have the correct number. First, if you have a goat that may have been microchipped previously, scan it to make sure it is not. Second, scan the microchip to verify the number. Third, scan the animal after implantation in order to verify and accurately record the number.

The microchip should never be inserted between the shoulder blades of a goat, for a variety of reasons. The FDA allows the use of microchips in animals that may end up in the food chain only if placed in areas not ordinarily used for human food, have little or no risk of migration to another part of the body, and are identified to establishments where they are taken for slaughter. If you microchip a goat, let buyers know, so they also can inform any future buyers of the fact that it is microchipped.

> **Tip:** The preferred site for microchipping goats is under the tail surface on the left side.

Three sites are indicated for microchipping of miniature goats. They are between the upper and lower hooves (dew claw), by the ear, and under the tail.

Dew Claw Injection Site. The recommended injection site is the right hind leg at the junction of a line drawn between the base of the dew claws and mid-line between the claws.

Secure the animal and its leg, squeeze the dew claws together to create a subcutaneous cavity ("tent" the skin) into which the device may be injected.

Prepare the site for an aseptic injection, by cleaning it with povidine-iodine (Betadine). Insert the needle subcutaneously (just under the skin) about ¾ inch and parallel with the long axis of the leg (toward the toe). Inject the device by depressing the plunger until it stops.

Release the external pressure on the dew claw as the needle is being retracted. Apply pressure at the injection site with the thumb to prevent the device from being drawn back through the cavity created by the withdrawal of the needle. Apply pressure for a few moments to prevent bleeding.

Ear Injection Site. The recommended injection site is the right ear located at the junction of the pole of the head and the crest of the ear (the front and top most point where the ear joins the head). Inject the device into the detectable depression extending forward superficial to the scutiform cartilage. (The scutiform cartilage is the depression felt when pushing the finger between the ear and the head.)

Adequately restrain the animal and secure the ear with one hand. Prepare the site for an aseptic injection. Insert the needle subcutaneously (just under the skin) in a downward and forward direction for about one inch (towards eye). Inject the device by depressing the plunger until it stops.

Retract the needle while applying pressure at the injection site with thumb to prevent device from being drawn back through the cavity created by withdrawing the needle. Apply pressure to prevent bleeding.

Under the Tail Surface. This is the preferred site for microchipping goats. The recommended injection site is on the under (ventral) surface of the tail about one-half inch away from the body proper. The implant is placed off the ventral midline of the tail (left side).

Adequately restrain the animal Prepare the site for an aseptic injection. Insert the needle subcutaneously (just under the skin) about one-half inch from the body of the animal. The 12 gauge needle should be advanced beneath the skin such that the implant is placed [at a distance of] the length of the needle from the insertion point. Inject the device by depressing the plunger until it stops.

With a fingertip, lightly apply pressure at the injection site where the transponder is located in order to prevent device from being drawn back through the cavity created by the withdrawal of the needle. Apply pressure for a few moments to prevent bleeding.

Tattooing

How to Tattoo. Insert the correct letters or numbers in the tattoo outfit. Impress it on a sheet of paper to assure that you have the letters or numbers in the correct order.

Hold, or have someone hold, the goat securely. You may need to tie a larger goat to help immobilize it. I have tried tattooing in the stanchion, but find it hard to hold the head securely.

Cleanse the inside of the ear or tail web with alcohol to remove dirt and wax. Dry the area to be tattooed, and smear tattoo ink on it. Make sure to avoid the veins, placing the tattoo characters between them.

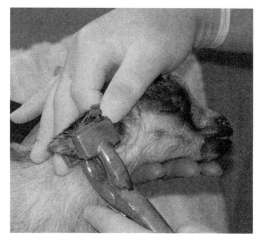

Quickly and firmly close the tattoo outfit and immediately release. Smear a bit more ink into the ear and rub for about 15 seconds. Some people use a toothbrush, but I have not found that necessary.

The wound will take one to three weeks to heal. In dark-skinned animals, holding a flashlight at the back of the ear will help you to read the tattoo.

Breeders who have been in the business for a while will probably continue tattooing; they have the equipment and the experience. As more registries accept microchipping, though, it may become more popular as an identification method. Each of us will have to make that decision for ourselves.

From *Ruminations* #33

A Word about Milk Goiter

Some goat kids develop a swelling on one or both sides of the neck, near the thyroid. The swelling, which is actually the thymus gland, usually grows during the first few months of life and eventually disappears.

It is believed to be caused by drinking the mother's rich milk (although it has been found in kids on milk replacer). My first experience with it was with a fat little kid whose mother wouldn't wean her. It lasted for months.

Goat owners and veterinarians who are unfamiliar with this phenomenon may think the goat has an abscess or an iodine deficiency, conditions with other obvious symptoms. Milk goiters are not harmful. Because the thymus gland produces cells that provide immunity in the goat, the swelling may actually indicate development of the immune system.

Goat keepers who are unsure about whether a neck swelling is milk goiter or not should consult a veterinarian.

For more detailed information on and a drawing of a milk goiter, see www.goatbiology.com/milkgoiter.html.

Coccidiosis
by Stacy Morris

Kidding season is over and you have a lot of cute, happy kids bouncing around. Then you notice that one has a little diarrhea, and treat it with some Pepto Bismol. That kid recovers, but then gradually every kid starts to develop the problem. While there may be another cause, the most likely culprit is coccidiosis, which will run rampant through a kid population if left unchecked.

What Is Coccidiosis?

Coccidiosis is a disease of the intestinal tract caused by coccidia, a protozoal parasite. The disease spreads from one animal to another by contact with infected feces, and is most severe in young, old or weak animals.

The parasitic one-celled organism that causes the disease coccidiosis is from the same class (Sporozoa) as the organisms that cause malaria. Parasites from this class form spores at some stage of their life cycle. The life cycle involves two stages. An asexual stage or schizogony and a sexual stage called sporogony. In the case of malaria, the schizogony stage takes place in man and the sporogony stage in the mosquito. The genus Eimeria undergoes all of its stages in one host, in this case the goat. Almost all species of animals have their own strain of coccidia. The coccidia of other animals, for example, rabbits and chickens, do not infect goats. The coccidia of sheep may be responsible for some problems in goats.

Coccidia are almost always present in a goat's environment. Here in the Pacific Northwest we have an ongoing struggle with coccidiosis due to the wet and moderate climate that is a perfect environment for it, while those of you in drier desert areas may never have to worry about it. When goats are infected with these parasites in small numbers, the coccidia cause very little damage and no disease. Adult goats have them, but are usually strong enough to resist them. People think of coccidiosis as a kid "disease" because kids have not built up a strong enough immune system to resist the coccidia yet; this is why kids show more problems with coccidia. You need to be concerned about and act on an overload of coccidia or worms.

Just after kidding, nursery pens and surrounding areas may be heavily contaminated with oocysts from does stressed from kidding. One- to six-month-old kids living in these areas are at the greatest risk. Shipping, ration changes, crowding stress, introducing new animals, mixing young with older animals and severe weather can trigger major outbreaks, because of stress and lowered resistance of the new and younger goats. Because occurrence of coccidiosis under these management systems often becomes so predictable, anti-coccidials can be administered preventively for 28 consecutive days beginning a few days after kids are introduced into the environment. Those who bottle-raise their kids should consider keeping their kids separate from the adult does and if possible rotating the area the kids live in annually. This should help keep the kids from being infected so quickly or severely and makes it easier to medicate them when needed.

The time from exposure to the coccidia in feces to onset of the illness takes from 5–13 days. The primary sign of coccidiosis is diarrhea with or without mucus or blood, leading to dehydration, emaciation, weakness, anorexia and death soon following, if not treated. Some goats are actually constipated and die without diarrhea. A full-blown infection attacks the intestinal lining, causing much discomfort and blood loss. Serious clinical infections can cause intestinal ulceration and scarring which will leave your goat stunted due to poor digestion and nutritional malabsorption. In addition, there have been several reports of hepatobiliary (having to do with the liver) coccidiosis with liver failure in dairy goats.

Diagnosis of intestinal coccidiosis is based on finding oocysts of the pathogenic species in diarrheal feces, usually at tens of thousands to millions per gram of feces. Finding oocyst counts as high as 70,000 in kids without overt disease is not unusual, but they may have problems with weight gain. In my experience any animal that is not growing properly and looks thin and unthrifty usually has coccidiosis, and doesn't usually develop diarrhea.

How Is It Spread?

Because feces spread coccidia, practicing strict sanitation is important. Coccidia is a very tough organism that can survive most disinfectants and harsh freezing cold weather conditions.

Proper management is key to treatment and prevention. Keep the kid's living area as clean and dry as possible, keep food and water dishes clean, make sure that food and water is supplied in such a way that the kids cannot stand in and soil it, and clip hair from doe's udders if kids are nursing. Flies can mechanically carry coccidia from one place to another; therefore insect control is also very important in preventing coccidiosis.

How Is It Treated?

Fortunately coccidiosis is treatable, although most goat keepers who have experienced an outbreak prefer to start their kids on a regular preventive program. Several drugs are proven to be effective for this purpose.

Lasalocid (Bovatech)—Not for use in lactating does, 15–70 mg/head/day, depending on body weight.

Monensin can be added to the feed at 20 g/ton (allowed in lactating dairy cattle now as part of a TMR only).

Deccox-M is a medicated powder (active ingredient: 3632 mg decoquinate/lb) added to kids' milk for 28 days. Some breeders swear by this regime.

Save-A-Kid milk replacer (Milk Products, Inc.) is a kid milk replacer with Deccox added. Several lamb creep feeds and mineral/salt mixes include one of the above ingredients; a similar product is Show Goat Express by Purina, which is made especially for goats. I have had fairly good luck using several of these products, but remember that these anticoccidials are preventives and are not effective if the animal is actually sick with coccidiosis.

Hoegger Goat Supply recommends using Di-Methox 40% (a concentrated IV Solution) in a prevention program. Starting at three weeks of age, mix ½ ml of Di-Methox with the kids' milk twice a day for one week. After that, continue to give ½ ml once a week until the kids are weaned.

Others use a similar treatment schedule (treat at three, six and nine weeks) with other drugs (Albon, Sulmet or Di-Methox).

Check with your vet and other local breeders to see what is working in your climate and area. If one doesn't work for you and you are sure you have coccidiosis, try a different regime.

Once coccidiosis has been diagnosed, early treatment can reduce the severity. The following have been effectively used for treatment:

> ## Tannin-containing Plants for Coccidia
>
> Tannin-containing plants are the focus of a study on Korean native goats that were infected with coccidia. The goats were fed condensed tannin-containing plants and for ten days after the coccidia eggs were studied. They found the pine needles and oak leaves to have rapid anticoccidial effects, with the number of coccidia eggs declining within two days by 40–44%.
>
> Note that goats tend to avoid tannin-containing plants, and that feeding them may be a problem and something that you should not undertake on your own, as they can be considered toxic.

Sulfaquinoxaline—for both treatment and prevention; may be used to treat affected kids in drinking water at 0.015% concentration for 3–5 days.

Sulfadimethoxine—Do not add to water or milk. Drench orally. Treat for five days at the following dosages: Day one: 1 ml per 5 lb. Days two thru five: 1 ml per 10 lb.

- Albon Concentrated Solution 12.5%
- Albon Soluble Powder 107 g pkg
- Di-Methox Concentrated Solution 12.5%—do not mix with water. Administer/drench directly into mouth.

- Di-Methox Soluble Powder 107 g pkg—dissolve one package (107 g) in three cups of water. Keep refrigerated. Administer/drench directly into mouth.

Sodium Sulfamethazine—Do not add to water or milk. Drench orally. Treat for five days at the following dosages: Day one: 1 ml per 5 lb. Days two thru five: 1 ml per 10 lb.

Sulmet Drinking Water Solution 12.5%—general broad-spectrum liquid, no mixing.

Aureomycin Sulmet Soluble Powder—Mix with water and treat for 28 days.

Corid (amprolium)—Over-the-counter product for preventing and eliminating coccidia. Comes in granular packets and gallon liquid. Use the gallon liquid and maintain better control over dosages.

Rule of Thumb: For prevention of coccidia, use 2 oz per 15 gallons of water; for treatment, use 3 oz Corid per 15 gallons of water. Limit the goats' water supply to one source and treat for five consecutive days.

> **Tip:** The use of Corid requires that the goat receive supplemental thiamine, to avoid polioencephalomalacia (goat polio).

According to the *Merck Manual,* amprolium has poor activity against some of the Eimera family, so it may not be the best choice available.

Other Considerations

Not all cases of diarrhea in kids are caused by coccidiosis. Usually, when we see diarrhea in a kid, we treat right away for coccidiosis, but you should always keep your eyes on the kid and be aware that the illness may be caused by something else. Salmonella, giardia, cryptosporidiosis, worm overload, *E. coli*, and enterotoxemia all have diarrhea as one of the major symptoms, so knowing what you are dealing with is vital before you start treating with an inappropriate medication. A good relationship with a vet is always recommended.

Although this article is aimed at coccidiosis in kids, adults can and will get coccidiosis. If you have adult animals with mysterious symptoms (anemia, weakness, unthriftiness and possibly diarrhea) that don't respond to your usual treatments, have them checked for coccidiosis. Usually you will find it is a combination of coccidiosis and a worm overload. Treatments are much the same as for the kids, but you will have to be more careful with the drugs you use if the does are being milked for human consumption. Most of the drugs do not have a recommended milk withdrawal time, so work with your vet and do a little research if treating coccidiosis in your adult herd.

From *Ruminations* #52

Health issues

Bloat

Gas occurs naturally in the rumen as a by-product of digestion. Bloat occurs when too much gas is trapped in the rumen and the goat is unable to release it by belching. Bloat can be life-threatening.

Bloating can be caused by overeating of lush pasture, overeating grain or when something, such as a large piece of apple or other fruit, lodges in the throat and blocks it so the goat cannot belch.

The signs of bloat include a bulging left flank, teeth grinding (indicator of pain), loss of appetite and lethargy. In the worst cases the goat will be down.

The treatment will depend on the severity of bloat, as well as the type, such as frothy or dry bloat. I have had success with oil drenches when I knew the animal had been into grain. For minor cases, kaolin pectin or baking soda will do the trick, as well as some good roughage such as straw. In these minor cases, massage is a good addition to treatments.

In more severe cases, the goat will need to have a stomach tube inserted and/or see a veterinarian.

An excellent discussion of treatment for the various types and degrees of bloat can be found on Maxine Kinne's web site at http://kinne.net/bloat.htm.

From *Ruminations* #55

Brucellosis

Brucellosis is uncommon in the US; the last case found in a goat was in 1999 in south Texas. The disease is, however, fairly common in goats in Mexico, as well as in other parts of the world.

Two brucellosis bacteria may affect goats—*B. melitensis* and *B. abortus*. The bacteria infect the reproductive organs and cause abortions and sterility. They also can pass to humans when shed in milk, urine, and other body fluids. Goats are particularly susceptible to *B. melitensis;* they are less frequently affected by *B. abortus*, which is predominantly a disease of cattle.

Signs of *B. melitensis* in goats include abortions in late pregnancy, retained placentas, weak kids, failure to thrive in kids, mastitis and decreased milk production. In some cases, goats may show no signs, or only some of them may abort their kids. Even if they successfully kid, they can still shed bacteria in body fluids, including milk. Those without symptoms are more likely to spread the disease because they aren't considered sick.

> **Tip:** When handling the placenta or fetus of a goat that has aborted, always wear gloves and make sure to wash your hands immediately after. This will help to prevent the spread of disease that may be the cause of the abortion.

Any time a goat keeper has goats aborting, it is essential to send a placenta to a lab to determine the cause. If only one case occurs, it may be an isolated incident, but if a pattern emerges, determining the cause is the key to preventing further incidents. Some people will freeze the placenta of the first goat that aborts, in the event that a pattern develops.

Brucellosis can be spread to people through raw milk, unpasteurized dairy products, infected goat meat and contact with secretions or aborted kids. Most cases in humans in the US come from raw milk products from Mexico. Pasteurization can stop the spread. In humans, the disease, also called Malta Fever, causes flu-like symptoms including fever, sweating, chills, headaches, joint pain, weakness, weight loss, nausea and depression. No safe, effective vaccine is currently available for people.

A vaccine to prevent *B. abortus* in cattle was licensed in 2003 and is now used in 49 states. A vaccine for *B. melitensis* in goats has been tested in Mexico and found to be effective for at least five years, but it is not used in the US. Recently, however, a test to detect *B. melitensis* bacteria in bulk goat milk was developed by a research chemist in the US, in the event that it becomes necessary.

Caprine Arthritis Encephalitis Virus

Caprine arthritis encephalitis virus (CAEV) is probably the most talked-about and contentious disease of goats, if not the most common. As one longtime goat owner explained to me, "in the 1980s goats were dying right and left from this disease, and it was horrible." Sounds like the AIDS epidemic.

As a more recent goat owner, I don't see the same thing. While acknowledging that, ideally, CAEV should be eliminated, I wonder whether—like HIV in people—the most susceptible individuals got sick and died and the more resistant ones survived and went on to reproduce, giving us the disease we have today—a majority of goats that don't even show any symptoms of their status. This is not to deny that those goats that do experience symptoms of CAEV may have long, drawn-out, expensive and sometimes painful deaths.

CAEV was first discovered in a goat herd in Washington in 1974. In that case, some of the kids had developed a progressive paralysis of their hind legs, along with a mild interstitial pneumonia. The disease was determined to be caused by a virus that infected the nervous system and named "viral leukoencephalomyelitis" (inflammation of the brain and spinal cord). While adults in the herd also had a chronic arthritis, the two diseases were not linked until a few years later, when the CAE virus (CAEV) was determined to be the cause of both.

CAEV is a lentivirus, in the same family as the human immunodeficiency virus (HIV), that infects goats and sheep. A similar virus originally found in sheep causes the disease Ovine Progressive Pneumonia (OPP), also known as Maedi-Visna virus (MVV). Both CAEV and OPPV/MVV frequently cross the species barrier between goats and sheep, so that now they are referred to collectively as small ruminant lentiviruses (SRLV). For the purposes of this article, I will use the term CAEV, since that is the term that most goat owners are familiar with.

> CAEV has five major clinical presentations: arthritis, encephalitis, interstitial pneumonia, mastitis, and progressive weight loss. A goat may exhibit one or more of these.

In the 1980s more than 80% of goats were determined to have antibody to (be infected with) CAEV. That number is now estimated to be around 10–30% of goats in the US. The majority of these goats (70–80%) show no clinical signs. However, having antibodies to CAEV does not mean that the goat has immunity. Once infected with the virus, the goat will stay infected for its lifetime and be a carrier even if it is not sick.

Illness Caused by CAEV

CAEV has five major clinical presentations: arthritis, encephalitis, interstitial pneumonia, mastitis, and progressive weight loss. A goat may exhibit one or more of these.

The arthritic form is the most common manifestation of the disease, affecting mainly those that are six months old or older. It is generally chronic, progressive, and painless, affecting the joints and synovial membrane (lining of the joints). The goat may appear stiff, favor one or more legs, lose weight, avoid getting up and walking, and stand abnormally. In more severe cases, the joints are swollen and painful. (CAEV has been called "big knee" because of this expression of the disease.) Studies have found a genetic predisposition to developing arthritis upon SRLV infection, which explains why goats in some herds never develop this manifestation.

The encephalitis form of CAEV infection usually affects kids between two and six months old, but can affect older goats. The kids act uncoordinated and may stand or walk oddly. Gradually they become paralyzed, usually in the back legs, and are unable to stand up. Sometimes they may be able to drag themselves by the front legs, and remain bright, alert, and responsive and do not have a fever. Because the disease is progressive, they eventually will develop further neurological symptoms, including depression, blindness, circling, tremor, difficulty swallowing, seizures and, ultimately, death.

Chronic interstitial pneumonia (inflammation of the connective tissue of the lung) is the form of CAEV infection that most commonly affects sheep, but also occurs in goats. The animal first develops a chronic cough, which leads to difficulty breathing, weight loss, increased rate of breathing and abnormal lung sounds. Enlarged lymph nodes may contribute to some of the respiratory distress.

Mastitis, especially interstitial mastitis (inflammation of the connective tissue of the udder), is another expression of CAEV. The udder becomes hard and distended, and often milk cannot be expressed. It often comes about around the time of kidding. Studies of milk production in goats with CAEV-caused mastitis have found conflicting results—from no reduction in milk production to 15% reduction.

The final major form of CAEV infection is chronic progressive weight loss. While wasting may be the only sign of CAEV, it often occurs with the other forms of the disease.

As previously mentioned, many goats will never show clinical illness from CAEV. Some of this may indicate the strain of virus that they have, but it also may have to do with whether they are well-fed, safe, live in a clean area and are not overcrowded. Stress can be a huge factor in bringing on clinical illness.

Transmission

CAEV lives in white blood cells and, consequently, can be spread through body fluids. It is transmitted from dams to their kids through colostrum and milk, which contain the virus. It also may be spread by long-term contact between seropositive and seronegative goats, through contaminated milk in milking equipment, or through blood on contaminated needles, clippers or dehorners.

It may be found in semen, but without an infection in the buck, breeding a positive buck to a negative doe is not considered problematic. However, a recent study found viral DNA in both pre-breeding and post-birth genital secretions of positive does.

There is still a question about transmission of the disease to kids during pregnancy and birth. An ongoing study showed 23% of kids of positive dams that were seronegative at birth and fed with heat-treated colostrum went on to test positive for CAEV. Investigation is ongoing.

Tip: Keep goats separate from sheep to avoid CAEV/OPPV.

Sheep and goats can be a source of CAEV or OPPV (MVV) transmission to each other. Close contact between the two species in crowded barns has been suggested as one mode of transmission. In experiments, lambs that nursed from infected goats became infected with CAEV. In one case, a goat was found to be naturally infected with a dual MVV and CAEV strain.

In Switzerland, where CAEV-caused disease was considered to be virtually eradicated in goats, a version was found that appeared more closely related to OPPV (MVV) than CAEV by blood testing in seronegative goats. This indicates that the sheep population is a potential source of infection for CAEV-free goats. The results of a recent epidemiological study support the conclusion that contact with seropositive sheep is the most significant risk factor for CAEV infection in CAEV-free goat herds.

Incubation Period

The incubation period is highly variable. Most goats become infected when they are very young, and develop disease months or years later. Encephalitis has been reported in a one-month-old kid.

The speed at which goats convert to positive may span a wide range—from a few weeks to several months, making a reliable blood test for CAEV quite difficult. Even with dams that were known to be positive, and exposure to the colostrum and milk, kids have been known to test negative more than seven months after birth.

According to the Washington Animal Diagnostic Disease Laboratory (WADDL), "These 'silently' infected animals test negative for antibody until the viral infection is activated by stress or other factors. It has not been determined whether these goats [are] infectious to other goats during the time they harbor the virus but remain seronegative."

A recent study found that antibodies to CAEV were not detected in 95% of kids (that were fed heat-treated colostrum) by 93 days after birth. All kids in that study were seronegative by 108 days, suggesting that goat keepers may begin testing kids at 3 to 3.2 months of age instead of the previously recommended four to six months.

Diagnosis

Besides a diagnosis based on clinical symptoms, blood testing is the easiest and most common method to determine whether a goat has CAEV. Two blood tests are available in the US: the Agar gel immunodiffusion (AGID) and the enzyme-linked immunosorbent assay (ELISA).

The AGID is the prescribed test for international trade. The ELISA is considered to be the most reliable of the two tests and is recommended by most goat vets and goat breeders. Both tests are about as reliable, but neither is 100% accurate.

In addition to the ELISA and the AGID, several methods based on the polymerase chain reaction (PCR) have been developed to detect CAEV through DNA testing; they are used mainly to detect positive status in animals that tested negative with ELISA and/or AGID. This test is not yet widely available, but in a 2008 study it was found to detect CAEV in four of 34 goats that had tested negative by AGID. The authors noted that "PCR may be a useful tool for decreasing the risk of breeding AGID false negative animals (CAEV carriers)."

A true positive result from any of these tests means that the goat is infected with CAEV. A negative result means that the goat either is not infected, or has been exposed but is not producing enough antibody to be detected. If there is reason to believe that the goat may be infected, it should be re-tested.

All seronegative goats should be re-tested every six months, unless they live in a closed herd (NO new goats introduced, no exposure to sheep [unless in the same closed herd] and goats not exposed to goats from outside the herd).

Treatment

No cure exists for CAEV and, to date, no vaccine has been developed. When a goat does develop symptoms, the treatment is for that problem. However, because CAEV is a chronic disease, treatment may be ineffective and the best that can be done is to keep the goat comfortable or euthanize it.

Strategies for Dealing with CAEV

There are several strategies for dealing with CAEV in a goat herd, and goat keepers may elect to use one or a mix of them. The first strategy is simply to avoid buying infected goats. Get a written contract for the sale, if you are unsure about the seller. Ask the seller whether she tests her goats, whether she has had CAEV in the herd, and if so, how she manages it. To be extra sure, quarantine a new goat for 30 days and have it tested for CAEV yourself.

If a recently purchased goat tests positive, ask the seller to take it back and refund your money. This is one area where a contract is useful.

Some individuals either euthanize or sell at auction any goat that tests positive for CAEV. Euthanasia seems rather extreme to me, considering that the animal has a good chance of never getting sick. But for some, even that small risk may be too much. For others, having to snatch the kids and rear them separately also may be too time- or resource-intensive.

If a decision is made to keep the animal(s), they should be allowed to live as stress-free a life as possible, and be properly fed and cared for. If a goat keeper has more than a few goats and breeds them, a couple of options exist.

First, if no animals leave the farm (for instance, the goats are freshened and extra kids are butchered), the kids can stay with their dams and CAEV serological status need not be addressed. If finances permit, testing twice a year or annually would be interesting simply for informal research purposes to see whether or when they become positive and/or whether positive goats exhibit any signs of the disease. If some goats in the herd stay seronegative, tracking whether their kids are also CAEV-negative would give some information on potential resistance to the disease.

> **Tip:** Test for CAEV twice a year initially, followed by annual testing with testing before kidding. Quarantine new goats and test twice, 30 days apart, prior to introducing them to the negative goats.
>
> In herds with both positive and negative animals, test negative animals more frequently to adjust the milking order so that negative animals are milked first.

The second option is to separate the kids from the seropositive mothers immediately after birth, raise them in a separate area on heat-treated colostrum and pasteurized milk or a substitute, such as safe (free of Johne's, mycoplasma or other transmissible diseases) goat or cow colostrum and routinely test all CAEV-negative goats. Make sure that CAEV-negative goats have NO direct contact with the CAEV-positive goats or sheep. This is the strategy of choice for breeders who plan to take their goats off the farm or sell them, as well as for those who want to eliminate CAEV from their herd.

Heat-treating Colostrum: Heat in a double boiler to between 133° and 138 ° F. (56–59° C.) and hold at that temperature for one hour. This may be frozen in pints (or less) for later use. Do not heat higher than 140° F.

Because of the risk of intrauterine or birth-related infection, even a goat raised under the second option is not 100% safe from getting the disease. This is one reason that routine testing is so important. If a kid tests positive after being pulled from its dam and fed heat-treated colostrum and pasteurized milk, it should be moved into the CAEV-positive herd or culled.

A CAEV-negative herd is worth working toward. With no CAEV in the herd, dams can raise their kids, giving them colostrum to boost their immune systems, raw mother's milk for the best nutrition possible, and mother-to-kid contact so they can bond and learn to act like goats. Goat keepers can also feel secure in not worrying about one more health issue for their goats and not inadvertently selling a CAEV-positive kid.

Caseous Lymphadenitis (CLA)

Caseous lymphadenitis (CLA), also known as pseudotuberculosis, is a chronic disease caused by the bacterium *Corynbacterium pseudotuberculosis*. It causes abscesses of both internal and external lymph nodes. In goats, external abscesses are more common, while sheep are more likely to have infection of internal lymph nodes, although often animals have both types.

Transmission

C. pseudotuberculosis enters the body through small wounds or mucous membranes and then localizes in the lymph nodes. It can take two to six months or more for an abscess to appear. The disease is then spread when the abscess bursts and another goat has contact with the purulent material. The bacteria can survive on and be spread by fencing, clippers, straw, hay and wood.

The internal form of the disease can spread to lungs and the digestive system from the lymph nodes. It often leads to wasting, but also can affect milk production and fertility or lead to mastitis, coughing, respiratory problems and neurological problems.

External abscesses are most commonly found under the jaw, around the mouth, in the neck, shoulder, upper leg and knee area. These abscesses can be differentiated from abscesses caused by *staph* or other bacteria because the pus has a cheesy texture. It is initially pale green and very thick, but can also be yellow or white.

> **Tip:** Abscesses are common in goats and usually are *not* CLA but are bacteria, such as *staph*, which has entered a wound from a splinter or other cause.
>
> As with CLA abscesses, goats should be isolated to avoid spread and the abscess lanced and drained.

CLA can generally be diagnosed by the location of the abscesses and the odorless, cheesy appearance of the pus. A blood test is also available.

Treatment

Antibiotics are not effective against the disease. Animals that are suspected of having CLA should be isolated from the herd. Abscesses should be lanced and drained, which will prevent them from spontaneously rupturing and contaminating the environment. Goat keepers should always wear gloves when treating these and other abscesses and then dispose of infected materials and disinfect equipment used in the treatment.

Some goat keepers have reported success treating CLA abscesses by injecting formaldehyde or formalin. There are two problems with this approach, however: Use of formaldehyde or formalin in food animals is prohibited, and this method does nothing for the internal abscesses that may still exist.

Prevention

Colorado Serum Company makes two CLA vaccines that are licensed in sheep. These are Case-Bac, which is a combination bacterin/toxoid, and Caseous D-T, which also contains tetanus toxoid and *Clostridium perfringens* type D toxoid.

Although some people use them extralabel in goats, the vaccines are not labeled for use in goats due to safety issues that came up during testing. These included large swellings at injection sites and post-vaccination lameness lasting up to a month. The company reports on their web site that they have learned that "a fair percentage of vaccinated goats will develop a fever and become lethargic for a period of days. These goats will sometimes go off feed or have a reduction of feed intake. Milking does can have a decrease in milk production. Vaccinating pregnant animals can increase the risk factors." They also noted that vaccinating goats that already have CLA will not solve the problem, but will make any reactions worse. As a result, they do not recommend the vaccines for use in goats.

Another option, if financially viable, is to obtain a definitive diagnosis from a vet and have an autogenous vaccine made from a sample of a CLA positive abscess.

As with other diseases, a good way to prevent CLA in a goat herd is to not bring it in. When considering purchasing goats, ask the breeders whether they have had a problem with abscesses, and carefully examine their herd. Ask whether they have had CLA in the herd, and if so, what they have done about it. These steps will go a long way in avoiding the introduction of CLA to a herd or purchasing an animal with hidden CLA.

Caprine Chlamydiosis
by Cheryl K. Smith

A goat herd owner has completed all breedings for the season and expects to provide proper nutrition and care while waiting an uneventful five months for kidding to begin. Then a few weeks prior to scheduled kidding, the worst happens, as does begin aborting (miscarrying) their kids.

One abortion may just be a fluke, but when more than one occurs, something is most likely wrong. Goat owners who fail to have the cause investigated are likely to experience more losses over time.

When an abortion occurs in a goat, keep the fetus(es) and membranes for laboratory analysis, isolate the doe for further examination and to ensure that others are not infected, and make sure to wear gloves and bury or burn aborted materials. Pregnant women are advised not to handle such products, as their own babies may be affected.

Chlamydophila abortus

One common infectious cause of abortion in goats is *Chlamydophila abortus* (formerly called *Chlamydia psittaci*). It was first reported in Germany in 1959, and then diagnosed in the US, UK, India, Japan and other parts of the world. In many areas, it is second only to brucellosis as the cause of infectious abortions, and is the most common cause in areas where brucellosis has been controlled.

Spread of Infection

The infection is spread through either shedding in feces, which is then ingested by other goats, shedding from the vagina two weeks before and after the abortion. Some goats may be carriers—either kids that survive their mother's infection or animals that are new to the herd. Often the abortion is the only sign that a doe has chlamydiosis.

Bucks may become infected when servicing infected does. While no studies have been done on goats, rams have been shown to become infertile and sterile as a result of genital infections caused by chlamydiosis.

Signs of Caprine Chlamydiosis

Unlike the disease in sheep, in goats chlamydiosis can cause an abortion at any stage of pregnancy. In addition to abortion, stillbirth or premature births of weak kids in the last month of pregnancy, in some cases other symptoms are present. These include persistent cough and breathlessness, pneumonia, keratoconjunctivitis (pinkeye), arthritis, and intestinal infections without clinical signs.

In some cases the goat kidding will have a retained placenta or uterine infection. Usually a goat that aborts recovers quickly and may have some brownish discharge.

The incubation period after exposure is sometimes only two weeks.

In herds new to chlamydiosis, the rate of abortion is often severe, with up to 90% abortions and a decrease in milk production. This lasts for about two to three years, with eventual immunity. Most goats will not abort a second time. No studies have been done to determine how often chronic infections occur in goat herds.

The only sure way to determine whether goats are infected with Chlamydophila abortus is laboratory analysis.

Effects on Humans

A number of cases have been reported in which people were infected through exposure to goats with chlamydiosis. In Germany during 2000/2001 kidding season, a woman in her 20th week of pregnancy developed a severe infection and aborted her baby. She had contact with aborting goats prior to that time; they were later shown to be positive for *Chlamydophila abortus*.

Preventing Abortion: Give 20 mg/kg oxytetracycline (LA-200 or Biomycin) IM at 105 and 120 days of pregnancy.

A French case was reported in 1991 in which a woman had a spontaneous abortion at 32 weeks following contact with a goat herd. That report noted that nine cases of maternal-fetal infection were known to be caused by contact with infected ewes. This was the first case in that country related to goats.

In the same year in the Netherlands, a woman whose husband worked with a goat herd where *Chlamydophila abortus* was known to exist had a preterm stillbirth of her baby. She initially developed an infection with rapidly worsening flu-like symptoms and eventually recovered after treatment with antibiotics.

Also of concern is drinking of unpasteurized milk by pregnant women.

Treatment and Control

Tetracycline is the treatment of choice. It can affect the replication of *Chlamydia* and can prevent abortions. The injection of 20 mg/kg of oxytetracycline intramuscularly (IM) at 105 and 120 days of pregnancy will prevent the abortion but not the shedding of the organism at kidding.

Live vaccines also are available, but can take up to three years to stop all abortions, because of latent infections in goats that were infected before being vaccinated.

Other measures that are recommended include isolation of does two weeks before and after kidding, culling kids from infected mothers, disposal of aborted material and disinfection of the area with a chlorine bleach solution.

Conclusion

Caprine chlamydiosis in a goat herd can seem like a disastrous situation. It can affect sales, showing, milk production and mortality. It can have far-reaching effects in other areas, as well.

The good news is that it is time-limited for an individual goat herd, as the goats develop resistance to it.

Sources:

1. Matthews, J. 1999. *Diseases of the Goat,* 2nd ed. UK: Blackwell Science.
2. Rodolakis, A. 2001. "Caprine Chlamydiosis" in *Recent Advances in Goat Diseases,* Tempesta, M., Ed. Ithaca, New York: International Veterinary Information Service. www.ivis.org
3. Nietfeld, J.C. 2001. Chlamydial infections in small ruminants. *Vet Clin North Am Food Anim Pract* 17(2): 301–14.
4. Pospischil, A., et al. 2002. Abortion in humans caused by Chlamydophila abortus. *Schweiz Arch Tierheilkd* 144(9): 463–66.
5. Berthier, M., et al. 1991. Materno-fetal infection by Chlamydia psittaci transmitted by a goat: a new zoonosis? *Bull Soc Pathol Exot* 84(5 pt 5): 590–96.
6. Meijer, A., et al. 2004. Chlamydophila abortus infection in a pregnant woman associated with indirect contact with infected goats. *Eur J Clin Microbiol & Infect Dis* 23(6): 487–90.

From *Ruminations* #53

Copper Deficiency in Dairy Goats

Copper deficiency in grazing livestock has been recognized in most developed countries, especially across Europe and North America as well as in Australia (pioneering work was done in Australia in the 1930s). As far back as the 1930s, localized cases of copper deficiency were discovered in Florida, the Netherlands, New Zealand and parts of Australia.

Severe primary copper (Cu) deficiency has been identified as a problem in dairy goats for a number of years now. While first identified in California large dairy breeds, it also been confirmed in Nigerian Dwarves, Boer, and Pygmy goats in Arizona, Colorado, Florida, Indiana, Missouri, Ohio, Oregon, South Dakota, Texas, Virginia, Washington, Wisconsin, and most of the New England states.

What causes copper deficiency?

Copper deficiency in goats is generally caused by low levels of copper in soil and feed. It may also be brought on by high levels of copper antagonists in the diet and from an excess of molybdenum in the soil. In addition, as in cattle, genetic differences in goat breeds and bloodlines within breeds are believed to affect the susceptibility of goats to both deficiency and toxicity. Research in cattle indicates that because some breeds of cattle absorb copper from the small intestine less efficiently, they need up to one and a half times the amount of copper as other breeds.

Copper's availability is reduced by iron, sulfur, molybdenum and zinc. The interaction between zinc and copper may be alleviated to a certain extent by maintaining the ratio of zinc to copper between three-to-one and five-to-one.

Sandy soils have traditionally shown deficiencies, but high organic matter soils, degraded black soils, wooded calcareous and grey-wooded soils can also be severely deficient. Copper deficiency may occur when animals graze on soils deficient in copper or soils with high molybdenum levels (+2 ppm). Copper intake should be five to eight times the molybdenum intake. Other causative factors may include pastures with high sulfate levels (+0.35% total sulfur), iron exceeding 250–300 ppm, or some combination of these.

Soil copper is in two forms, $Cu2+$ and $Cu(OH)+$. In general, the plant availability of copper in the soil decreases with an increase in pH of the soil. As pH rises absorption increases and the solubility of the oxides decreases. Deficiencies can occur naturally in soils that are naturally high in pH or have been over-limed. The opposite can occur in very acid soils. This is true of all the micro nutrients except selenium and molybdenum. Studies have found that only 35% of pastures in the US have adequate copper levels. However, producers should be aware that copper content of pastures may vary from spring to fall.

Water, usually from a well or hot springs, may have substantial amounts of sulfur, which reduces the availability of copper. Water from alkaline soils is more often high in sulfur.

While some believe that heavy parasitism causes a loss of copper in goats, recent research with copper boluses (in sheep) have shown that they reduce parasitism. This may be a "chicken or egg" situation. Other factors that have been implicated include Johne's disease and excess zinc supplementation.

Results of 1999–2000 USDA-APHIS-VS-NAHMS (USDA's National Animal Health Monitoring Service) studies indicated that in general, the Western states have lower mean serum copper concentrations compared to other regions. The mean serum copper concentration for operations in the Western regions was 0.63 ppm, while the Midwest and Southern regions recorded 0.70 ppm. In the US 28.7% to 57.8% of pastures had molybdenum (Mo) and iron (Fe) levels high enough to cause copper malabsorption. To this can be added malabsorption through excessive sulfur intake. Alfalfa is notorious as a crop that is susceptible to copper deficiency. Wheat, barley and oats can also be deficient.

Soil-applied copper will generally have long-lasting residual effects. Beneficial effects from 1.3 to 2.7 pounds of copper per acre have persisted undiminished for up to 35 years (western Australia). Copper can be applied as organic compounds in the form of CuEDTA, copper ligninsulfonates, and copper polyflavonoids.

Why is copper important?

Copper is necessary for the absorption and utilization of iron, it helps oxidize vitamin C and it works in conjunction with vitamin C to form elastin, a chief component of muscle. It also helps with the formation of red blood cells and bone structure. A copper deficiency prevents the bone marrow cells from reaching maturity.

Copper is actively transported through the intestinal wall and stored in the liver. Copper deficiency prevents iron from being incorporated in hemoglobin, resulting in anemia, which is indistinguishable from iron deficiency. Copper plays a role in iron absorption and mobilization. Copper deficiency impairs the formation of connective tissue proteins, collagen and elastin. Weak bones (osteoporosis), and defective arterial walls are the more obvious manifestations.

What are the signs of copper deficiency?

Copper deficiency has been shown to cause a variety of problems from immune deficiencies to thin, rough and faded coats to stillborn kids. In the 1990s, a goat herd in Arizona lost all but three kids to copper deficiency.

In California, when copper deficiency was first identified in goats, herds with the most severe problems suffered immeasurable genetic losses. On the other hand, many herds did not have specific or classic copper deficiency symptoms, but were plagued with a multitude of miscellaneous conditions, such as frequent *staph* lesions on the udder, nose, mouth, and chin (occasionally the entire body) and bald tail tips or light spots on the nose. They also had frequent minor maladies such as pinkeye and ringworm.

Some herds had serious problems with increased cases of mastitis, including more than a dozen cases of gangrene mastitis in a two-year period. They also experienced ruptures of the uterus and pre-pubic tendons (abdominal wall hernia), huge hematomas (blood clots) following injections or minor injuries, osteoporosis, twisting or bending of the front legs and/or feet in kids and pregnant yearlings, anemia, and the list goes on. Strangely some herds experienced NO obvious problems yet lab work showed these herds to have comparable deficiency.

Young kids are most often and most severely affected, with everything from the classic symptoms of ataxia caused by demyelination of the spinal cord (a breakdown of the insulating fatty coverings that surround the nerves in the brain and spine) to light colored rings around the eyes, thin hair over the nose and/or around the eyes and/or ears, small size, general weakness or sore joints and general failure to thrive.

Deficient does are not able to get kids on the ground with adequate levels of copper to maintain them in good health. Often they are so extremely deficient that they suffer from osteoporosis (soft, porous bones that bend and fracture easily), severe anemia, or other health problems. Some are unable to survive at birth, some appear normal at birth with symptoms showing soon after or even weeks/months later depending on the level of the deficiency and the individual animal.

How is copper deficiency diagnosed?

While liver biopsy is the most reliable test of the true copper status of the animal, few private practice or University-affiliated veterinarians are willing to take the risk of performing liver biopsies on breeding stock. In the last few years some have shown interest in the procedure and successful caprine liver biopsies have been done at Texas A&M. The University of California at Davis has developed a procedure for performing liver biopsies on cattle that is proving successful. For now though, no reliable test on a live animal is readily available to the small goat keeper; obtaining a copper level from the liver of a deceased animal is still the only accurate indication of the Cu status that we have.

Blood testing for copper is a poor second choice, and we seldom test the blood levels anymore. Copper is carried in the blood in a variety of ways and conventional blood tests measure only the total copper content. Of this, usually only about 3% is available for use in enzymes. For various reasons, the blood level does not correlate to the level in the liver. The blood level can be normal, low, or even elevated, while copper stores in the liver and kidneys are extremely deficient.

While liver copper and total blood copper are used alone as indicators of copper status, these do not take into account the correction of the symptoms of clinical copper deficiency. The efficacy of copper supplementation of ruminants is the ability to correct the symptoms of clinical deficiency and should not be judged by the supplement's ability to raise the copper content of the body. Copper levels in hair samples are highly variable.

Tip: Work with a veterinarian, if possible, before using large amounts of supplemental copper. Copper toxicity has been reported in some areas where goats were given Copasure boluses or minerals formulated for beef cattle.

How is copper deficiency treated?

If copper nutrition was as simple as determining the copper levels in the base diet and adding a highly available copper source/supplementation, copper deficiency would not be a problem. However, because copper absorption and metabolism can be affected by molybdenum, sulfur, calcium, zinc, iron, manganese, cobalt, lead, cadmium, and selenium, deciding how much supplemental copper is required is not always easy.

Goat breeders in California tried oral supplementation of different mineral mixes high in copper (up to 1100 ppm) and feeding of other than goat specific feeds (horse pellets, horse minerals, etc.) to correct the problem. To date none of them has succeeded in bringing up the body stores of copper. However, other areas of the US have had excellent results with just the addition of a mineral mix high in copper, such as one for cattle or horses. Note: Mineral mixes labeled for sheep AND goats do NOT contain adequate copper for goats. Generally goats should not be fed sheep minerals without some other form of copper supplementation.

Absorption of copper can vary from zero to as high as 75% depending on a number of factors. Copper availability in most feedstuffs fed to farm animals is between 1% and 15%. Most minerals contain copper oxide in powder form, but availability is poor when provided this way. The mineral passes through the gut with little absorption.

Breeders in Southern California, where the deficiency was determined to be especially severe, have found that using copper boluses (capsules of copper oxide wire) dosed to weight is the most effective means of raising the liver copper levels to within normal limits in goats. The first boluses were brought into the US from New Zealand in the spring of 1994; since then cattle copper boluses that are downsized to goat doses have been available.

Thousands of goats have been on these boluses since the 1990s. Continuous laboratory work on bolused animals indicates that we are achieving normal liver concentrations of copper. No cases of copper toxicity have been found and only one elevated liver copper level was diagnosed. Liver concentrations remain in the low normal (30–80 ppm) except in rare cases.

Bolus need to be administered at five- to six-month intervals to maintain adequate body stores. After about four months, liver stores start to fall rapidly. In order to best protect the neonatal kids, the boluses are best used at times that will keep the doe's copper level up during her entire pregnancy.

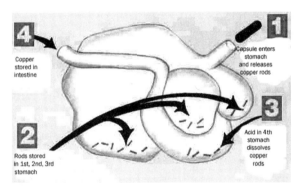

Recent Minnesota research with mice showed that brain development of the unborn kids was affected by copper concentration in the mother's diet. Some breeders are routinely giving boluses (0.625 to 1.35 grams) to kids early on (two to four weeks old), which has proven successful.

This is an ongoing program. We get additional/new information, ideas, etc., constantly. As time goes on the regime may change to less or more frequent bolusing or perhaps a completely different method of correcting the problem. The ultimate goal is to correct the primary source of the problem (hay/feed) so supplementation is no longer needed.

Both private and university affiliated veterinarians are interested in and working on this issue. Laboratory work has been done primarily by the California Animal Health & Food Safety Laboratory System (CAHFS/UC Davis) and the University of Arizona, with Texas A&M recently becoming involved with problems identified in that area. On the other hand, many veterinarians are still unaware of the problem and even argue against its existence, though as time goes by more professionals are seeing both the problem and the results and are recommending copper supplementation to goats in these areas.

How copper boluses work

Gelatin capsules containing copper oxide needles provide relatively long term protection against copper deficiency.

Copper oxide needles are brittle rods (1 to 8 mm long, and 0.5 +/- 0.1 mm in diameter) made by oxidizing fine copper wire. They are nontoxic when given orally, and they can be given in doses sufficient to establish long-lasting reserves of copper in the liver. Their properties were discovered by Australian scientists, who found that a combination of small particle size and high specific gravity (2.0 and 7.0) cause them to become trapped in the folds of the abomasum.

The gelatin capsules contain thousands of tiny, blunt copper oxide rods. When given orally, the capsules dissolve in the rumen, releasing the copper oxide rods, which then lodge in the abomasum. There they release copper for the animal's immediate requirements and reserves. The rods dissolve completely over a period of time.

Dosage: Dose goats at the rate of 1 g copper oxide in bolus form per 22 lb at five- to six-month intervals.

Laboratory work has shown that liver and kidney concentrations start to fall rapidly after about four months.

The copper oxide particles remain in the folds of the abomasum for at least 32 days. The excretion rate of copper from the copper oxide particles has been shown to be about 0.2 grams by weight per day, which allows for the safe absorption without apparent toxicity. The accumulated liver stores of the absorbed copper can protect the animal against copper deficiency for up to six months.

To be effective the copper particles must be administered by a conventional balling gun which delivers the capsule direct into the gullet where it is swallowed. The rods should not be chewed, as this may cause the particles to pass on through the animal in greater amounts than intended.

To get the most out of a copper bolusing program, make sure that parasite load is under control in the goats.

Adapted from an article by Joyce Lazzaro, in *Ruminations* #50

Tyzzer's Disease and Copper Deficiency

by Deborah Niemann Boehle

More than a month has gone by since my daughter complained, "Muse is drying up!" I told her that was impossible since Muse had only freshened two months earlier. A goat does not dry up after two months fresh. Something must be wrong, I told her.

For a couple of months I had been concerned about Muse, our only adult Lamancha doe who lives with our herd of about 20 Nigerian Dwarves. She didn't shed her winter coat completely. She had patches of dry hair hanging on until we finally shaved her in mid June. She also seemed too thin to me, but a couple of standard goat breeders told me that meant she was just putting all of her energy towards making milk. I had dewormed her with morantel tartrate pellets when she kidded in May and, not liking the way she looked, I dewormed her with ivermectin in June. A couple of days after her production started to drastically drop, we noticed that she had diarrhea, so we started her on Di-Methox for coccidia.

Then on a Thursday night at the end of July, my younger daughter came running into the house screaming that Muse was having convulsions. My husband ran out to the pasture and carried her into the barn. Her body was ice cold, and she appeared to be blind. We put her under a heat lamp, and I gave her a shot of oxytetracycline, assuming listeriosis or some type of disease that would be treated with an antibiotic. We obviously didn't have any time to spare. My daughters, my husband and I sat there brainstorming diseases and symptoms, and nothing fit. A half hour later, at 11 pm, she was dead.

The next morning, I was on the phone with the vet, talking about a necropsy. About 10 days later, he called me with the initial results—Tyzzer's disease. He said there was no known cause, no known mode of transmission and no known cure. He also said he had never seen a case of it in a goat.

I googled Tyzzer's, and learned that it is a rodent disease. It is most common among wildlife such as muskrats and rabbits, but also occurs in mice and other small rodents, such as gerbils and hamsters. Then I used the library database at Illinois State University. I put "Tyzzer's" in one search field and "goats" in another and got zero hits, meaning that probably no recorded case of Tyzzer's in a goat exists. The university database includes hundreds of professional journals, including those read by veterinarians and other medical professionals. I found case studies of Tyzzer's in a cockatoo and in a conure (parrot), as well as one in a panda bear, but nothing in a goat. A few cases have also been reported in foals and calves. One thing all non-rodent victims have in common is either a compromised immune system or a young age.

I also found a case study of a rabbitry in Canada that had an outbreak of Tyzzer's. They lost half of their rabbits in a few days. After they put oxytetracycline in the drinking water, they didn't lose any additional rabbits, leading the researchers to conclude that oxytetracycline was a cure. However, when one reads about cases of wild animals contracting Tyzzer's, it seems that it usually only wipes out about half of the population in an area, although no one explained why that might happen. Since it is a disease that occurs mainly in wild rodents, scientists have little incentive to conduct research on the disease.

Tyzzer's is a *Clostridium* (a gram positive bacterium related to the one that causes enterotoxemia), but it not one of the *Clostridia* for which goats are vaccinated, because goats are not rodents. No vaccine has been created for Tyzzer's, probably because no one has a financial incentive to create a vaccine for wild rodents—or even pet rodents.

We puzzled over how and why our goat could get a rodent disease, and a couple weeks after we received the initial necropsy results, the vet finally called with the rest of the lab work. Muse's copper liver level was 4.8 ppm. The normal range is 25–50 ppm. He didn't volunteer the copper level, and when I first asked he said it was a "little low." Then I asked for the exact level and gasped when he told me. Prior to the necropsy, when I had first asked for a copper level on her liver, he tried to talk me out of it. "She wouldn't have died from copper deficiency."

I insisted on the test, because I had suspected that her poor coat and poor body condition might have been a result of a copper deficiency. I had been reading about the problem on one of the Internet groups after a long-time breeder told me via e-mail that she thought one of my other goats might be deficient after I sent her a picture of the goat. The more I read about copper deficiency, the more I thought it was a problem for a few of my goats, but especially for my LaMancha.

When I went to buy my LaMancha two years ago, the breeder said that during her pregnancy, her legs had grown crooked, which he had attributed to "two big kids." Since her legs looked normal when I saw her, I thought nothing of it. Now I realize that this is a sign of copper deficiency. In the past few months, her hooves had been growing increasingly crooked, and hoof trimming didn't help. When I noticed "decreased libido" and "decreased fertility" on the list of symptoms for copper deficiency, I was reminded that she only came into heat twice the first year I had her, and she never got pregnant. This past year, she was finally pen-bred because I had difficulty determining whether she was in heat; she didn't get pregnant until December.

Being so severely deficient in copper would explain why she would contract a disease of another species, but we were still wondering how she could get it. We have a livestock guardian dog who doesn't let anything in the pasture, which would rule out rabbits and muskrats. Then one day recently I scooped out some dewormer pellets and saw mouse droppings in them. The light bulb went on, as I recalled giving the pellets to Muse after she kidded.

Obviously Muse contracted a rodent disease because she was nutritionally compromised. I've spoken to three different vets about copper deficiency in goats, and all three denied that it could be a problem. Even faced with the lab results, the vets did not recommend anything other than switching to a brand of goat feed and minerals that has more copper than the feed I was using. Two of the vets told me that my goats were not deficient in copper without even letting me explain the symptoms I had seen in my goats. Copper deficiency can be a real problem, and like many issues with goats, breeders have to educate themselves and make their own decisions about herd management.

From *Ruminations* #58

Coughing Caprines
by Jill Landers

Cough, cough, cough…. "Honey, a couple of the does are coughing. They weren't coughing yesterday, I wonder what's wrong?"

Sound familiar? If you've had goats for any length of time, chances are at least one of them has coughed! If you show, chances are that all of them have coughed at one time or another. Coughing can be as simple as a bit of dust in the lungs, to a serious upper respiratory infection.

How are we supposed to figure out just what's causing the cough and then what do we do about it? As a herdsman, it is our job to know our goats when they are in good health. What is "normal" for each goat? When we come to know our goats when they are at their best, it is much easier to determine when they are sick or "not quite right."

Tip: If the goat has a deep-seated pneumonia or bronchial infection, the cough may be moist and feeble.

The first thing I do when I notice a goat coughing is watch them. Are they acting "normal," i.e., eating, cudding, fighting with pen mates? Then I try to determine when the cough occurs. Do they cough when eating? After being butted by somebody in the herd? Randomly?

What is the weather like? Some goats do have allergies! I'm not one to run for antibiotics right away, so unless they are showing signs of infection (more on this later) I will wait a day or two, while carefully watching them.

The nature of the cough itself can tell you a lot. According *Goat Medicine,* by Smith and Sherman, if the irritation lies in the upper respiratory tract, the cough is typically dry and powerful. If the goat has a deep-seated pneumonia or bronchial infection, the cough may be moist and feeble.

A goat may cough for numerous reasons. Common causes may include:

Butting
Sometimes when the bully butts smaller goats they will cough, just like humans do when the wind gets knocked out of them! If this is a frequent occurrence, you might consider penning smaller animals away from the mature ones.

Fencing
We use cattle panels a lot on our farm. We were noticing several does randomly coughing and figured out that they would stick their heads so far through that it irritated their tracheas and caused them to cough.

Collars

Have you checked your goats' collars lately? Kids are especially vulnerable to the ever-tightening collar that will press on their throats and cause coughing. Be sure you can get at least two fingers easily between the throat and the collar.

Dust

Have you switched feed? Does your feed have a cloud of grain dust that billows up when you scoop it? Just like in people, dust can cause a cough in goats. Try to limit the dustier grains like rolled barley, alfalfa pellets (some are dustier than others), etc., or you can have your feed mill add a little molasses to bind the dust.

Allergies

I had a goat that would cough and sneeze during the high pollen seasons. Is the mold count high? Is the car windshield yellow from pollen? Try a dose of Benadryl (2–3 ml) for very young kids to 7–10 ml for adult animals and see if the coughing subsides. If they quit, you can assume an allergen is causing problems and either give Benadryl regularly, or just know that it's not a communicable cough.

Worms

I will also look at the worming schedule and rotation. Lungworms will cause a non-specific cough. According to my vet, liver flukes and many other worms at various stages are cough inducers too. Are your goats due to be wormed? I will typically use something that will get lungworms and liver flukes. Ivermectin (0.13 mg/lb (0.3 mg/kg)) and Valbazen (4.5 mg/lb (10 mg/kg)) are good options (label states not to administer during first 45 days of pregnancy—check with your veterinarian).

Foreign Body

Sometimes a goat will get a piece of grain stuck in its throat. This seems to happen to the gluttons in our herd, you know, the ones that have to be first in line and then try to cram as much in their mouths per bite as possible! Most of the time they will cough it up. Occasionally a piece of grain will be stuck in their sinuses. This will result in a nasal discharge, and sneezing and coughing. Contact your vet for assistance.

"Fair Cough"

These bugs are the nasties we get at various shows and fairs. As you walk through the fair barn, if you hear several goats coughing, you can almost assume that you get to bring something home with you too…don't you feel lucky? There are several things I do to ward off the fair coughs (or at least make them a little easier to combat):

First, I use a concoction of herbs/immune stimulants (if you don't use natural medicine, skip this part!). I also give my does 500 mg of vitamin C per day for two weeks preceding the show and three times a day at the show and a couple weeks following. They love the chewables and I give them to each doe while I am milking. I will also give each animal a good dose of probiotic before we leave home. Some other breeders will feed garlic (I can't get mine to eat it).

Trick: To help prevent transmission of diseases to goats at shows, fair or other events, spray the pens and the ground with a 1:4 bleach solution.

Second, when we arrive at a show we always spray the pens our goats will be in with a 1:4 bleach solution. I also spray the ground. According to our vet, bleach solution is one of the best disinfectants, and he recommends spraying pens (especially at fairs where other animals have been in them) to avoid many diseases.

Third, when we are at a show, I will do my best to isolate my animals from any others. Sometimes this means hanging a tarp between someone else's goats and my own. It's not that the other person's goats are necessarily sick, but each herd seems to have its own "local bugs" that it is immune to, and I just want to try to avoid any additional stressors on my girls. Yes, you may still get the fair cough, as sometimes it's airborne, but a little extra trouble could save a lot of hassles later.

Fourth, when we get back home, I will use an antibiotic immediately on any animal that looks "off." Usually, a few days of Bio-mycin will clear up most fair coughs. Nuflor is also a good choice, especially if other goats in the barn were acting pretty ill. I also give a dose of B-Complex orally to boost overall well-being. We have had some show coughs that have lingered for 2–3 weeks.

Pneumonia and Respiratory Infections

A goat with pneumonia is usually a pitiful sight. Elevated temperature, labored breathing and general depression are all possible signs of pneumonia or another respiratory infection. This is a critical situation that needs medical attention immediately. I pull the goats out and do a more thorough examination.

Begin by taking their temperatures (normal goat temp is 101.5–103° F, but can be higher in hot weather or following fighting, chasing to catch, etc.), then respirations (10–30) and pulse (70–120).

If the temperature is up a degree or so, but the animal is looking and acting normal, I usually administer 1000–2000 mg of vitamin C twice a day for a day or two and see if that doesn't cut it. I will also give fresh Echinacea leaves as an immune booster (I grow it in my flower gardens). Then I will give a shot of B-Complex—I just learned that several breeders give this orally, which is a lot easier on the goat and me!

If they are acting "off" (not eating, off by themselves, breathing difficulties, not cudding, sometimes they will have a snotty nose, sometimes not) and their temperature is elevated, I will administer antibiotics. My favorites for respiratory problems are Bio-mycin (4.5 ml per 100 lb (9 mg/lb) SQ every 36 hours or 3 ml per 100 lb [6 mg/lb] once daily); or Nuflor (6 ml/100 SQ every four days, or 3 ml/100 lb IM every two days). Always remember to administer a probiotic daily (FastTrack, Probios or plain yogurt with active cultures work great). I will also continue with vitamin C to help boost the immune response. You can also use VetRx or Vicks VapoRub on their nose to assist breathing. I then isolate them, with a friend if possible, keep them warm, dry, offer water and hay and check them frequently. If they don't look a little perkier in 24 hours I contact my veterinarian for his input.

Goat keeping would be an easier endeavor if there were simple answers as to why they are coughing! Unfortunately, our goats haven't read all the books we've studied, and their symptoms aren't always easy to diagnose. So we have to be good goat detectives to determine what may be wrong with our dear caprine friends!

From *Ruminations* #30

Enterotoxemia

Enterotoxemia, also called "over-eating disease," is one of the most common diseases of goats. The cause of the disease is the toxin produced by the bacterium *Clostridium perfringens* type C or type D.

These bacteria can be found normally in soil and in goats' intestinal tracts. Enterotoxemia (literally, a condition of poisoning in the intestine) is a condition in which the bacteria grow out of control. It occurs when certain conditions allow this overgrowth and the movement of food through the intestinal tract is slowed. This may occur with overfeeding of grain, lush pasture, milk or milk replacer. It may occur more often when a goat is stressed or has another illness.

It is often seen when a kid is first exposed to grain or large amounts of milk. For this reason, kids should be gradually introduced to grain, and if they are dam-raised and removed from their mothers at night (usually to allow the owner to milk the mother, or for showing) and then nurse from a full udder. Intake of these large amounts of food slow the intestinal tract.

Type D *Cl. perfringens* is more common in kids, while Type C is more often seen in adults. Type D produces what is known as "epsilon toxin." It damages the blood vessels, and is absorbed very rapidly. This is why, in some cases, goats that die of this type of enterotoxemia sometimes show no signs of illness before they go down. Those that absorb only small amounts of toxin live longer, allowing for time to treat. Unfortunately, by the time symptoms develop, it may be too late, *unless the antitoxin was given at the first sign of a comprised gut.*

> **Tip:** Enterotoxemia can be avoided "by giving the antitoxin preventatively when the goat's gut is compromised" by bloat or another condition.
> — Sue Reith

Type C enterotoxemia produces "beta toxin," which leads to intestinal necrosis and severe intestinal hemorrhage and usually death. This type is found more often in adults.

In the peracute form, death can occur within minutes or 12 hours of signs of disease and generally affects kids.

Acute enterotoxemia usually affects kids, lasting 4–24 hours and usually ends in death. Symptoms include high temperature, severe abdominal pain, watery diarrhea, depression and going down. Kids with enterotoxemia often cry loudly. In some cases, the goat will have seizures or throw its head back. Treatment is generally ineffective and watching a goat with acute enterotoxemia is very distressing for the goat keeper.

Subacute enterotoxemia is more often found in older kids and adult goats. The illness may last for several days or weeks, and the goat loses its appetite and has intermittent diarrhea. Goats with the subacute form of the disease usually recover if they are treated and can begin eating again.

Chronic enterotoxemia is more likely to occur in adults and causes them to be sick intermittently for a number of weeks. They have loose feces, diminished appetite, look "off" and exhibit a drop in milk production.

> **Treatment:** 5 ml CDT Antitoxin SQ (for kids) and up to 40 ml for adults; Banamine; penicillin; Bo-Se, thiamine, Pepto Bismol, probiotics and baking soda, according to label dose.
>
> Decrease milk consumption by half to unweaned kids, followed at least an hour later with an electrolyte solution for the first 24–36 hours.
>
> Give NO GRAIN for at least a week after an episode of enterotoxemia.

While the symptoms discussed above may be helpful in diagnosis of enterotoxemia, they are also seen in salmonellosis, coccidiosis, intestinal torsion (twisting) or other conditions. Fecal examination and culture can help in differentiate these conditions. Often a determination cannot be made on necropsy.

Giving antitoxin, with noticeable improvement, points to enterotoxemia as the causative agent.

Vaccinate with *Cl perfringens* C, D and T (tetanus) toxoid and avoid the sudden introduction of large amounts of feed or milk to help to prevent the disease. Vaccinate previously unvaccinated adults two times, at 4–6 week intervals. Vaccinate again during the last month of each pregnancy to "boost" immunity and provide antibodies in colostrum for the immediate protection of the newborn kids. Vaccinate kids at two or three weeks of age and four to six weeks later.

Foot Rot

Foot rot requires the existence of two conditions: untrimmed hooves and the presence of a bacterium. Preventing the first condition is the key to controlling or preventing foot rot.

Foot rot is a condition in which the hard portion of the hoof separates from the underlying tissue, causing lameness. In the early stage, the skin between the hooves becomes inflamed (also known as foot scald.) If allowed to progress, the tissue at the heel or the inside of the toe starts to separate from the hoof, causing a bad smell. In the final stage, the entire sole and possibly the whole hoof wall separate from the inner tissue. Not each hoof progresses at the same rate, so the goat may be favoring one leg or the other.

Tip: Make sure to trim the hooves of goats without foot rot first and to soak hoof trimmers in a bleach solution between infected goats.

The organism that causes foot rot is the bacterium *Dichelobacter nodosus*. The bacteria can survive in soil for two to three weeks and in the hoof wall for several months. Foot rot is more common in wet or muddy conditions.

Foot rot is usually brought onto the farm by a new goat that is infected. It is then spread to the soil, where it can affect other goats. It also can be transmitted on the boots or shoes of people who visit the farm, which is one reason to require that visitors wear disposable boots or disinfect their shoes before they enter goat areas. Another way to prevent spread is to disinfect the feet of goats that are infected and then ensure that they don't go in areas that will infect other goats for two or three weeks.

The first step in treatment is to trim affected hooves, focusing on removing infected areas. The next step is to soak the foot for a few minutes in zinc or copper sulfate, such as Dr. Naylor Hoof 'n Heel or Koppertox. Goats with chronic foot rot should be treated with an antibiotic such as penicillin or tetracycline.

You also can purchase a goat "boot" that it can wear to keep the foot dry and also to retain medication for a longer period of time. This allows the foot to heal and provides a barrier against further infection.

Vigilance is important in eliminating or preventing foot rot on the farm. Routinely trim hooves and keep an eye on goats' movements to spot and deal with any lameness before it goes too far. Check the feet of any goat you plan to purchase and quarantine the goat for 30 days before introducing it to the herd.

Johne's Disease

Johne's (pronounced "yo-nees") disease is a contagious, slow-developing and fatal disease of the intestinal tract. It is also known as paratuberculosis. Johne's disease is caused by the bacterium *Mycobacterium avium* subspecies *paratuberculosis (M. paratuberculosis)*. It is related to the bacterium that causes tuberculosis and shares some characteristics of that disease. The bacterium can replicate only within the animal; it can survive in soil or water for over a year, but it cannot multiply.

The disease is hard to detect. Because of the long incubation period, it is rarely seen in kids. It also has vague signs. Goats, unlike cattle, do not get diarrhea from the disease, and the only symptom may be weight loss, while still eating well. Johne's disease also may be mistaken for gastrointestinal parasites, malnutrition or caseous lymphadenitis (CLA). Currently the prevalence in goats is not even known, although the disease is considered to affect goats worldwide.

Goats normally get Johne's disease when an apparently healthy, but infected, new goat enters the herd. The animal then contaminates the farm by shedding the bacteria in its feces. Other goats, particularly kids, then ingest the bacteria. Kids also can acquire it if the doe is shedding the bacteria into her milk or colostrum, or gets feces on her teats from unclean bedding.

Another route of exposure occurs in utero, when a goat is in the later stages of infection. At that time, if the animal is pregnant, the fetus can also become infected. The kid will appear healthy at birth, although spontaneous abortion of infected fetuses have been reported in cattle.

> **Tip:** Be cautious about using cow colostrum for newborn kids. Infected cows may shed *M. paratuberculosis* in their colostrum and milk. Many dairy herds in the US are known to be infected. Even pasteurizing the colostrum will not guarantee destruction of *M. paratuberculosis* (145 degrees F. for 30 minutes is not always effective). Some breeders recommend always using milk replacer rather than cow milk.

In some cases, kids get Johne's disease from drinking cow colostrum or milk. Goat keepers may be trying to prevent their kids from getting CAEV or another disease from their dam, and unknowingly expose them to another disease, such as Johne's, that can be spread between species.

Goat Health Care

The best way to avoid Johne's disease is to ensure that new goats are not infected. Most goat breeders do not currently test for the disease. Each goatkeeper will have to decide for him- or herself whether to purchase only tested goats, considering the risk of the disease and the costs that requiring such testing may entail.

Keeping feeders and water buckets clean will minimize the spread. Separating does into the clean, dry kidding pens can also help to prevent spread. Goats that have unusual weight loss or diarrhea should, of course, be quarantined and tested for likely diseases, including Johne's, by a reputable lab.

Johne's disease can be detected through fecal culture and blood test, although neither are completely accurate. If a goat dies from a slow wasting, consider having it necropsied to determine whether Johne's may have been the cause.

> **Tip:** Look closely at any herd that you plan to purchase goats from. What is the body condition of the herd in general? Of the animal you plan to purchase? Ask the owner about how many animals are culled and the reasons for culling. Check the body condition of the dam of kids you intend to purchase. If you have concerns, negotiate to have the dam tested for Johne's disease.

No effective treatment has been found for Johne's disease and it is not considered curable. Goat keepers will have to assess whether they can afford to pay for treatment that likely will only improve the goat's symptoms, and whether they can afford to separately house a sick goat that has no breeding potential and no chance of cure.

Those Pesky Lice!
by Cheryl K. Smith

You may have seen it: goats acting restless, scratching incessantly, sliding down the hill on their briskets, and sometimes even losing patches of hair, especially on their legs. When you check their coats, you find little bitty critters inhabiting them. These are lice, and they can cause intense irritation and itching, usually during the winter and in early spring.

Lice are wingless, flat insects that are very tiny in size. Two kinds of lice may affect goats: biting and sucking. The biting lice feed on dead cells on the skin surface and cause itching. Sucking lice suck blood from their host, but also cause itching. Severe infestations of these can lead to anemia and in extreme conditions, death due to blood loss. The scientific name for the biting louse that affects goats is *Bovicola capra*; the sucking louse is *Lithognasus stenopsis*.

Interestingly, different animals have different lice that have evolved specifically to them (although sheep and goat lice seem to affect the other). This means that goat lice will not affect humans and human lice will not affect goats. In addition, because the lice are dependent on the animal, if removed they will soon die. They spend all of their time on the host animal.

How do you know if a goat has lice? Diagnosing lice usually is not hard. You can often tell that lice are affecting the goat if patches of hair are missing or you see hair attached to fences, trees and other areas that can be rubbed against to relieve the itching. Since the lice may inflame the hair follicles, another sign of infestation may be a rough, or poor quality, coat.

If you look closely in good light or with a magnifying glass, you can see the lice and their eggs, called nits, attached to the hairs. They are more likely to congregate along the topline in goats, so check there first. The nits are grayish white and can be found in large numbers on the hairs next to the skin.

If you look at the lice under a microscope, you can determine whether the lice are the sucking or biting variety. A sucking louse has a head that is larger than the thorax (middle section of the body); a biting louse has a thorax that is larger than the head. You can also see that they have small claws to hold onto the hairs of the animal.

The nits hatch into larval lice, called nymphs. They then feed and molt three times before becoming adults. The whole life cycle takes about 30 days.

How do you prevent lice? The best control for lice is regular brushing. However, if you have a large number of goats, this may not be feasible. Healthy goats are also more likely to be able to keep their lice burden down, as well as not showing the secondary effects caused by these pesky critters.

Fresh air and rain are reputed to prevent lice. (I have yet to see the day that my does willingly go out in the rain.) Related to this is the fact that overcrowding can lead to more severe infestations, as the goats infect one another.

What should you do when you determine that a goat has lice? Often the lice are not severe enough to require treatment; in fact, when the weather warms up and they can be clipped for shows (or comfort), the lice will leave of their own volition. Sunny, hot weather and lack of a warm place to burrow makes the goat a less friendly home.

If the lice infestation is bad enough to require treatment, or summer is a long way off, a number of insecticides are recommended for treating goat lice. These insecticides may be applied by dipping the animal in a vat (not very practical or affordable for most of us); spraying; dusting; pour-on (spot-on or down the back); or by giving an anthelmintic orally or by injection (e.g., ivermectin).

For insecticidal treatment, I prefer using pyrethrins, because they are natural insecticides, produced by certain species of the chrysanthemum. The flowers are dried and the oils removed. The dusts that come out of this process are poisons that penetrate the nervous systems of insects. In some cases, other chemicals are added to provide a more toxic effect.

Pyrethrins aren't known to be stored in the animal's body or excreted in milk. They do have some risk, however. Use of dust with pyrethrins should be done in a well-ventilated area, with care taken not to inhale the powder.

Two other nice things about pyrethrins is that 1) they decompose and lose their toxicity through exposure to sunlight and air, and 2) they can be used on lactating goats, unlike other chemical treatments.

A number of oils and herbs have insecticidal properties. Most well-known of the insecticidal oils are tea tree and neem. Rosemary, lavender, lemon and geranium essential oils also have properties known to repel insects.

If your goats are healthy and minimally infested with lice, you can "ignore it and it will go away." Keep an eye on their behavior, the appearance of their coats, and look for any changes. This will tell you whether you need to treat for lice, or just wait until it is warm enough to clip them and let them go out in the sun for the truly natural cure.

From *Ruminations* #43

Listeriosis

Listeriosis, also known as circling disease, is caused by the coccobacillus *Listeria monocytogenes*. It is a bacterial infection that can survive under a wide range of climates. *L. monocytogenes* is found in soil and in the intestine and feces of a variety of mammals, as well as in contaminated sileage. When goats graze close to the ground, they can ingest the organism and further spread the bacteria. It is becomes a problem in the spring and winter, mainly in housed goats.

Treatment: 6 ml procaine penicillin per 100 lb, every six hours, along with dexamethasone and thiamine.

Listeriosis can affect both kids and adults, but is more common in adults. Some goats are carriers, exhibiting no signs of the disease, but potentially spreading it to other members of the herd.

The illness usually manifests as encephalitis (inflammation of the brain) in goats. They will become disoriented and depressed and lose their appetite. The facial muscles can become paralyzed and they tend to salivate a lot. Listeriosis also infects the uterus of goats, which can cause an inflammation of the placenta, infection and death of the fetus, abortion, stillbirth and death of the newborn kids.

L. monocytogenes also can be shed in milk for more than a month after the birth, leading to infection in humans who drink the milk or eat the milk products. Most reported cases of listeriosis in humans have been traced to cheese made from cow milk. In some cases the milk was even pasteurized.

Listeriosis is treated with high doses of procaine penicillin—every six hours, until 24 hours after symptom relief. Dexamethasone should be given to relieve any brain swelling. Some recommend the use of thiamine, as well. Make sure the goat is well-hydrated and well-nourished—tube feeding, if necessary. Recovery is possible if it is caught early and treated aggressively, but many cases in goats result in death within 24–48 hours of symptoms.

Mastitis

Mastitis is an inflammation of the mammary gland that may be due to bacteria, viruses, chemical or toxic insults or physical trauma. It is more likely to occur where poor milking hygiene and poor milking techniques occur, or when a goat's udder is injured in any way.

Mastitis often is "subclinical," with no signs or just a decrease in milk production. In these cases, testing, either of somatic cell counts (SCC) through a lab, or with a California Mastitis Test (CMT), can determine whether a goat has mastitis. (Note that goats have a higher number of somatic cells [white blood cells that appear in response to infection or injury] in their milk than cows, so cow values should not be used.) An SCC higher than 1,500,000/ml suggest intramammary infection. SCCs are highest during and right after milking, so testing should be done at the beginning of milking, when possible. In addition, they will tend to rise toward the end of lactation.

Tip: For does with an elevated CMT or SCC, but no signs of mastitis, take a milk sample, freeze and submit for culture in the event that clinical signs develop.

Mastitis also may be clinical (when actual signs can be observed in the udder or in the milk), manifested by sudden onset, redness, swelling, hardness, pain, grossly abnormal milk, and reduced milk yield. Signs of clinical mastitis can include lameness, one side of the udder larger than the other, hot or cold udder—indicates infection or lack of blood flow, anorexia, decreased milk production, abnormal looking milk.

Chronic mastitis is an inflammation that is ongoing. In some chronic cases it cannot be determined by looking at the udder or the milk, but can be diagnosed by cultures.

Common Causes

While the list of mastitis caused by viruses and bacteria is more extensive, the most common causes of mastitis in goats include *Staphylococcus aureus (S. aureus), E. coli, mycoplasma* and caprine arthritis encephalitis virus (CAEV).

S. aureus can lead to a chronic mastitis that varies in its presentation. The goat will have an increase in SCC and a decrease in milk production. This is one of the most common causes of gangrene mastitis, which can lead to eventual sloughing of the udder, or the need for mastectomy or euthanasia. Gangrene mastitis causes a watery, dark red secretion that may be accompanied by gas bubbles caused secondary infection an organism such as *Clostridium spp.*

S. aureus mastitis can show up prior to kidding or anytime during lactation. In addition to decreased milk production, the goat may lose her appetite, have an increased temperature, and grind her teeth in pain. In some studies the rate of death with this disease is 40%. Treatment can be intensive and even if treated the doe can become a carrier.

This mastitis is treated with intramammary antibiotics and a systemic antibiotic to fight the infection, as well as Banamine to reduce inflammation. Treatment is labor-intensive, requiring that the doe's udder be stripped every one or two hours, if possible.

> **Tip:** Blood in the milk is usually the result of the bursting of or injury to a blood vessel within the udder. It generally isn't mastitis....
> Glenda Plog
> Adare, Queensland, Australia

Coliform mastitis, caused by *E. coli*, is not common in goats. It can cause severe mastitis and make the doe very ill. It should be treated the same as *S. aureus*.

Mycoplasma mastitis is often under-diagnosed because often cultures are not done routinely and the incubation time for diagnosis is two weeks. Two types of mycoplasma mastitis that occur in dairy goats are *Mycoplasma putrefaciens* and *Mycoplasma mycoides spp. Mycoides*.

M. putrefaciens is a form of mastitis that is more common from late winter until early spring. It often affects both sides of the udder and is highly contagious. Signs include loss of appetite and depression in the doe, with foul-smelling, clotted milk. The goat may stop lactating entirely and the udder may temporarily atrophy. Does with this kind of mastitis must be isolated from the herd to avoid spreading.

> **Tip:** To avoid mycoplasma mastitis problems take a sample of colostrum from each doe at kidding for lab culture.

Symptoms of *M. mycoides spp. Mycoides* can resemble CAEV and some believe that the two have been confused. Both can spread through a herd. In addition to mastitis this type of mycoplasma is characterized by arthritis, meningitis, pneumonia, abortion and even sudden death. Goat keepers who see clinical mastitis in does and arthritis in nursing kids, or kids drinking unpasteurized milk should think of *M. mycoides spp. Mycoides*. If a new doe with this bacterium has been introduced to the herd, mastitis and abortions may be the signs, and milk should be cultured, as the most common cause of a herd outbreak is a carrier doe.

Does with either form of mycoplasma mastitis may need to be culled; at a minimum they should be milked last or with each other, using separate equipment.

CAEV is another common cause of mastitis. It may be subclinical, or may manifest as a hard, but not swollen, udder even in a first freshener. The goat generally has no loss of appetite or fever, but will have an increased SCC. Any doe that develops a hard udder with her first kidding should be tested for CAEV mastitis.

Testing and Treatment

When a goat has a high SCC that is of concern, take a milk sample to send to a lab for testing. If acute clinical mastitis develops, call a veterinarian for assistance, or when a vet is unavailable, begin treatment immediately. However, it is best to avoid routinely treating goats without first getting a lab culture of their milk. This will help to prevent the development of antibiotic resistant bacteria, as well as provide assistance in determining the appropriate treatment. Recommended cultures include aerobic, anaerobic and mycoplasma culture and sensitivity.

> **Tip:** Avoid routinely treating goats without first getting a lab culture of their milk.

Prevention

Preventing mastitis is the same as keeping the goat in good health, in general. Make sure the barn and milking area are clean and ventilated. Make sure fences are safe so goats don't get injured. Dehorn all goats and regularly trim hooves. Isolate sick goats.

Milking procedures and cleanliness also are important. Trim udder hair to prevent accumulated dirt and feces. Wash the teats and udder with a warm disinfectant solution before milking and dry each goat with a separate paper towel. Take one or two squirts of milk prior to milking for examination. Do a CMT once a month at the beginning of milking. Consider quarterly routine cultures of the entire herd's milk. Always milk goats with CAEV or other problems separate or last, if you choose to keep them in the herd.

Goat keepers use a variety post-milking teat dips, and some also give each goat an intrammary infusion at the end of the drying off period.

> A retrospective review looked at medical records of 17 goats and three cows that had radical mastectomies (removal of the udder) after severe mastitis. Two died, but the others did well. The authors concluded that it is a safe and effective procedure that can extend an animal's life.
>
> *J Am Vet Med Assoc* 231(7): 1098–103

Gastrointestinal Parasites

Gastrointestinal parasites can present a serious management problem in goats. Approximately 14 species of parasites (worms) can affect goats, although some are much more common than others, and many can be treated in the same way. An important point is that goats should not be worm-free; they—like all mammals—need a certain level of these parasites to keep their immune systems healthy. Only when worms overload the body do problems occur.

> Wet weather in spring and summer [is] associated with increased annual incidence of fatal gastrointestinal parasitism in goats.
>
> *J Am Vet Med Assoc* 231(7): 1098–103

A study at the University of Iowa in the 1990s showed that giving gastrointestinal worms to people with Crohn's disease—an autoimmune disorder in which the body fights against the colon—improved their symptoms.

The most common worms in goats are gastrointestinal nematodes, which live in the digestive tract—mainly the stomach and the small intestine. Other worms that are commonly found include the tapeworm, the meningeal worm (often found in deer), the lungworm and the liver fluke.

Excessive worms are caused by such things as a depressed immune system, consuming too many worm larva, filth and lack of sanitation, rainfall, close grazing, etc. Worms function in the ecosystem to keep animals from overrunning the ecosystem when production conditions are good and they also prevent all animals from starving when there is a shortage of food. One cannot eradicate worms on your farm; you have to learn how to live with them and use management to control them to levels which do not harm animal production.

One of the most problematic worms in the US (particularly the Southeastern US) is the barber pole worm *(Haemonchus contortus),* so named because it is red and white striped. This worm lives on the goat's blood and lays eggs prolifically—at a rate of 1000–6000 per day! Because of this fast rate of reproduction and the fact that it sucks blood, the way to tell that a goat is overloaded, or heading toward overload (and ultimate death) is to look for anemia. (See FAMACHA, page 30, for monitoring this parasite.) Other parasites can also cause anemia.

Another common sign of a goat that is overwhelmed by the barber pole worm is bottle jaw. Bottle jaw is a swelling below the jaw, and also can be caused by other parasites, such as the liver fluke.

The barber pole worm needs rain to infect small ruminants, so are less problematic in areas with limited rain. However, even in areas with a short rainy season, this parasite may develop dewormer resistance quickly, because a generation can be completed in only one month.

Two species of gastrointestinal parasites that are problematic in areas with mild temperatures are the bankrupt worm *(Trichostrongylus colubriformis)* and the brown stomach worm (*Telodorsagia circumcincta,* previously called *Ostertagia).* Although not as lethal to goats as the barber pole worm, these parasites are a problem in the fall and winter. They commonly cause diarrhea, rough coat and general unthriftiness.

The tapeworm rarely kills goats, but can cause young goats to develop poorly, get a pot belly and sometimes diarrhea. Tapeworms are easy to identify because sections of them drop off in the feces and look like rice grains. They live in the small intestine, absorbing their food from digestive tract. In cases where a goat's immune system is down, tape worms can grow and affect the goat's nutritional status.

The tapeworm life cycle involves an insect called a grass mite, which eats the eggs. They develop in the mite and the goat eats the developed egg from the grass, where the mite lives. After 40 days, segments of the tapeworm will appear in the feces. Cold winter weather can stop the cycle in pastures. Otherwise it may take a full year for the pasture to be clean of this parasite. Valbazen is the drug of choice for tapeworms.

The meningeal worm is another worm that creates problems in wet weather. It can cause partial paralysis in goats. It is a disease of deer, who spread it to goats through feces. The larvae in the feces are eaten by snails and slugs, and then go through a three or four week period of development. When grazing goats inadvertently eat the slugs or snails they become infected. The larvae, once inside the intestine, migrate to the spinal cord—a process that takes about 10 days. During migration to the brain, tissue is destroyed, which is what causes the paralysis. Other neurological signs also occur, including circling, abnormal head position, blindness, and difficulty walking.

> **Tip:** The Drenchrite® test can be done to determine which dewormers will be effective. This test is offered through The University of Georgia's Parasitology lab. For a cost of about $250, resistance patterns to all dewormers can be determined with one test.

Meningeal worm is usually a problem in autumn and winter. Treatment is often ineffective, although it may be treated with high doses of fenbendazole and ivermectin, as well as steroids.

Prevention requires keeping deer away from areas used by goats, whether by fencing or the use of guardian dogs. In addition, goats should be kept out of wet, swampy areas, compost or leaf piles. Many goat keepers have ducks, geese or guinea hens to eat slugs and snails.

The liver fluke is another worm that uses snails in its lifespan. Goats get them from eating grass that contain larvae that have matured in the snail. The flukes end up in the bile duct, where they lay eggs that are then released into the environment through feces. These flukes invade the liver and cause the goat to bleed internally.

Signs of an acute infection with liver flukes include anemia, loss of appetite, "going down" and then dying. If a goat has the chronic form of the disease—with fewer worms—it loses weight, eats poorly, develops a rough coat anemia and a rapid heart rate, and sometimes bottle jaw.

Liver flukes are commonly a problem in the winter and spring because they depend on a snail that is dormant at other times of the year. The choice of dewormer for liver flukes is Chlorsulon or Valbazen. Only Chlorsulon is effective against immature flukes and mature ones.

Lungworms usually occur during cooler months. Some are spread when the larvae are coughed up or released in the feces and then mature. Hot weather and hard freezes will kill them, but they can thrive in cool, wet conditions. Other lungworms use snails or slugs in their development cycle.

Lungworms can cause painful breathing, chronic cough, unthriftiness and, ultimately, death. A special fecal exam, called a Baermann, is needed to find lungworm larvae.

Tip: Proper administration of dewormers. All dewormers, other than Moxidectin, should be given orally and should be delivered with a drench gun over the back of the tongue into the throat.

Treatment is more effective if the goat is kept off feed for 24 hours prior to drenching.

Coccidia are another gastrointestinal parasite. They mainly affect kids, and are discussed in depth in "Coccidiosis" in the Kid Care chapter.

Parasites should be regularly monitored and controlled to avoid illness and spread to other goats. Monitoring includes routine observation and fecal testing (see Parasite Control, p. 33), as well as FAMACHA.

Control of parasites encompasses keeping barns clean, pasture management, monitoring parasite levels and anemia level, and deworming, as necessary. Because knowledge of appropriate parasite control is still evolving, goat keepers should also work closely with their vets and keep up to date on the most effective methods.

Pinkeye

Pinkeye, or keratoconjunctivitis, usually starts with a runny, squinty eye. The inflammation may increase, making the eye red and swollen (hence the name pinkeye). If isn't treated or it doesn't resolve on its own, eventually the cornea will start to become hazy. The eye may turn cloudy and then opaque, and the goat may begin to lose its eyesight.

In the worst cases, the goat may become blind or develop ulcers on the cornea. These can lead to systemic infection, which ultimately may kill the goat. In some cases of *Chlamydia*-caused pinkeye, the herd may also experience related abortions.

> **Tip:** Keep on hand terramycin ophthalmic ointment and treat pinkeye three or four times a day until it resolves. Sometimes only one or two treatments are necessary.

While often mild and self-limited, goat keepers should monitor goats with pinkeye to ensure that it isn't progressing, and involve a veterinarian if the initial treatment is ineffective.

Goats can have two types of pinkeye: infectious and non-infectious. Non-infectious pinkeye can be caused by vitamin A deficiency, sunlight, hay dust or another irritant. It will not spread to other goats, although several goats may have it at the same time, due to exposure to the same irritant.

Infectious pinkeye is caused by a virus or bacterium—e.g., *Chlamydia* or *Mycoplasma*. It can be brought on by stress, immune system problems or stresses (such as CAEV or other illness), nutritional deficiencies or extreme weather. It then can be spread from goat to goat—often by flies.

> Injectable oxytetracycline 200 mg/ml (LA-200 or equivalent), given at 5 cc/100 lb for 3–5 days, may be used SQ, in addition to topical eye ointments.

Most cases of pinkeye resolve by themselves. In severe or long-standing cases, the goat should be separated from the herd and treated. Isolation has the added benefits of ensuring that the sick animal is not being bullied and made sicker, and has access to food and water so it is better able to maintain good nutrition. Because sunlight can aggravate the condition, ideally the goat should be isolated in a cool, dark place.

Always wear gloves when treating for pinkeye or other contagious diseases. Treat and feed the sick animals after the healthy animals are fed.

The first line of attack is to flush the eyes with sterile saline. If the pinkeye is caused by an irritant, this may be all the treatment necessary. The most effective treatments for infectious pinkeye, when the eye has not ulcerated, are terramycin ophthalmic ointment, or a drop of port wine or oxytetracycline in the affected eye. In severe cases, consider treating with injectable oxytetracycline may be indicated in addition to the topical treatment.

Triple antibiotic ophthalmic ointment can be used on eyes that have ulcerated. This is a relatively expensive veterinarian prescription item. *Do not use steroid medications on ulcerated eyes.*

If the pinkeye progresses and fails to respond to treatment, the goat will need to see a vet, who can prescribe stronger medications, perform diagnostic testing or even do surgery, if necessary.

Pneumonia

Pneumonia is a common illness in goats. It can occur in both kids and adults, and can have both infectious and non-infectious causes—including parasites, CAEV, CLA, sudden change in the weather, viruses, poor nutrition or dietary changes, stress of transport and poor ventilation.

Pneumonia is an inflammation of the lungs that can come about when bacterial, viral, or parasitic infections, or other irritants breach the normal tissue barriers that protect the lungs.

Signs

Signs of pneumonia include a fever with a temperature of 104–107 degrees F., a moist painful cough, difficulty breathing, discharge from the nose or eyes, loss of appetite and depression. Crackles can be heard when listening to the lungs of a goat with pneumonia.

> **Tip:** Naxcel is approved by the FDA specifically for treatment of pneumonia caused by *Mannheimia haemolytica* and *Pasteurella multocida* in goats.
>
> Administer 3–4 cc/100 lb SQ daily. ½ cc for kids. Repeat for three to five days.

Types of Pneumonia

The most common type of pneumonia is caused by a bacterial infection. Healthy goats normally have bacteria in their lungs. *Mannheimia haemolytica* and *Pasteurella multocida* are two of these. Newborn kids are exposed to these bacteria, but usually do not develop pneumonia if they receive colostrum, which contains antibodies, are in a clean, uncrowded and ventilated environment, are not stressed and do not have another problem (such as a virus or lungworms) that can impair the lungs.

Colorado Serum Company has a vaccine against pneumonia caused by *P. multocida* or *M. haemolytica*. It is given at a dosage of 2 ml SQ twice, two weeks apart.

Viruses, such as parainfluenza-3 (PI3), are common in goats and can increase susceptibility to infection by inflaming the respiratory tract. The lentiviruses that cause ovine progressive pneumonia and caprine arthritis encephalitis also can cause a chronic pneumonia in goats.

Other organisms, including Mycoplasma and lungworms, can cause lung problems that lead to pneumonia. Lungworms are easily treated, although if they are severe the goat may be left with a chronic cough.

Several species of mycoplasma can cause pneumonia and contagious caprine pleuropneumonia in goats. Concurrent signs may include inflammation of the joints, the udder and the eye. The drug of choice for treating mycoplasma-related pneumonia is tylosin, or Tylan.

Pneumonia in Kids

Young kids tend to be prone to pneumonia. Kids that develop pneumonia can lose weight, become lethargic, have a runny nose, rapid breathing and/or coughing, and develop a moderate fever. They often stop nursing adequately. If the pneumonia is not detected right away, the lungs can be damaged and treatment less effective. If the kid is treated and recovers, it may be more likely to have further bouts with pneumonia, develop a chronic cough, or not grow well.

> **Trick:** To lower high fever from pneumonia, you can give a young kid ¼–½ baby aspirin. Keep in mind that fever is the body's way of fighting infection; but a very high fever can be dangerous.

Treatment

Regardless of the cause of pneumonia, keeping the goat hydrated is essential. If it is not nursing or is unwilling or unable to drink, it can be tube fed. In severe cases, SQ injections of sterile water will help hydrate the animal. Banamine or another anti-inflammatory drug can be given to reduce fever and pain, and antihistamines can be given for congestion.

> **Treating lungworms:** Administer Valbazen at 2X the cattle dose for three consecutive days, or Ivermectin at 2X the cattle dose for three consecutive days. Repeat in 14 days.

If the cause is bacterial, rather than viral, a course of antibiotics should be given. In serious outbreaks, you may want to treat all exposed animals with a therapeutic dose of antibiotics for several days.

If lungworms are the underlying cause of the pneumonia, deworm the goat with an anthelmintic that kills lungworms and treat other goats that are coughing, as well.

Prevention

Pneumonia can be prevented by providing proper housing and nutrition, plenty of space and attention to cleanliness.

It can be caught early if the goat is regularly observed and found to be acting depressed or "off." The first step is to take that goat's temperature.

If a goat does develop pneumonia, isolate it from the others to prevent it from spreading. Use separate buckets and feeding equipment and clean everything well after the animal has recovered.

If a goat develops a cough, follow up to determine what might be causing it. (See "Coughing Caprines," p. 127.)

Polioencephalomalacia (Goat Polio)

by Stacy Morris

Goat Polio (also known as polioencephalomalacia or PEM) is a disease characterized by a disturbance of the central nervous system. The brain of an infected animal becomes inflamed and swollen, and eventually becomes necrotic (starts to die). PEM is a metabolic disease often seen in goats that are raised in crowded conditions, although it may affect goats in smaller herds as well.

It is caused by a thiamine (vitamin B1) deficiency. Goats must have an adequate supply of thiamine in order to break down and use glucose. If thiamine is either not present or present in an altered form (thiaminase), then brain cells die and severe neurological symptoms appear.

Causes of PEM

PEM sometimes occurs in goats that are on high-grain diets, and diets that include plants high on thiaminases and sulfur. Thiaminases also are found in a few plants—including as bracken fern—and certain fish and shellfish. Other causes of thiamine deficiency include feeding moldy hay or grain, overdosing with amprolium (Corid) when treating coccidiosis, feeding molasses-based grains (horse feeds), sudden changes in diet, the dietary stress of weaning, excess sulphur in the diet and reactions to some dewormers. Each of these can interfere with vitamin B1 production.

In a case reported earlier this year in the *Journal of Veterinary Medicine,* a number of cattle on a ranch developed PEM after eating an excessive amount of barley malt sprouts. The cause was determined to be excess sulphur.

Using antibiotics destroys flora in the rumen, which can cause an imbalance and can lead to thiamine deficiency. The gut must be repopulated with live bacteria (e.g., with Probios) after using antibiotics or scour medications.

Normally ruminants are fairly resistant to thiamine deficiency since microbes in the rumen provide them with enough thiamine. However, when they eat bracken fern or other plants with thiaminases they may develop PEM.

Grain-rich diets can encourage the growth of certain thiaminase-producing bacteria in the rumen. These bacteria, including *Clostridium sporogenes* and a few species of *Bacillus* can produce enough thiaminases to cause thiamin deficiency. The brain and heart, which require large amounts of thiamine to function properly are affected first by thiamine-deficiency.

What PEM Looks Like

An increase in PEM tends to occur during winter, when goats have less access to high quality hay and forage, and owners start feeding increasing amounts of grain prior to kidding. It usually occurs in kids or young goats and without warning.

Affected animals will stand or sit alone, appear to be blind and sometimes arch their necks back and stare upwards, in what is known as "star-gazing." The medical name for this is opisthotonus. They can also become disoriented, stagger around, lose their appetite and do not want to drink. Temperature and respiratory rate are usually normal but the heart rate may be depressed. They may initially act excited, but then become quiet and dull. Normally only a few individuals out of a herd are affected.

An animal afflicted with PEM may go down on its side with its head thrown back. Its legs may be rigidly extended and convulsions may occur. Animals with PEM will often press their heads against a wall or post. Initial symptoms can look like enterotoxemia (overeating disease) or listeriosis. A component of "overeating" may actually be involved, in that the rumen flora have been compromised, or become unbalanced. As the disease progresses, convulsions and high fever occur, and if untreated, the goat generally dies within 24–72 hours.

Diagnosis and Treatment

PEM can be diagnosed through laboratory testing, but because the condition is so fast-acting, the owner does not have the luxury of the time that such tests take. The only effective therapy is thiamine, and treatment can result in improvement in as little as two hours, if the disease is caught early enough. In fact, some recommend giving thiamine when these symptoms are present—if the goat improves, it may "prove" the diagnosis. In any event, no harm will be done by trying it since thiamine is water-soluble and quickly eliminated from the body through the kidneys.

Thiamine is a veterinary prescription but it is inexpensive. Anyone with goats should keep thiamine on hand. The dosage is related to body weight; Initially, intravenous (IV) dosage is best, but if no one is available with the skills to give an IV, subcutaneous (SQ) or intramuscular (IM) can be used. Some goat owners even give thiamine orally after the initial treatment.

If thiamine is unavailable but injectable B-complex is available, the recommended dosage is 12 ml every 8 hours. This is because the treatment is based on the amount of thiamine contained in the vitamins. The key to overcoming PEM is early diagnosis and treatment. Complete recovery is possible under such circumstances.

Dexamethasone is often administered along with thiamine to reduce brain swelling. Although recovery is usually quick, if significant brain damage has occurred, the recovered animal rarely regains a satisfactory level of productivity.

Prevention

Some authorities recommend giving thiamine to the other goats in the herd as a preventive when a case of PEM is diagnosed in one. Drinking water should be tested for sulfur and a different source used if it is found to have a high level. Animals that have recovered from PEM should be introduced to grain diets in gradual steps to avoid a sudden increase in thiaminases-producing bacteria in the rumen. In addition, feeding management needs to be reviewed and evaluated.

Pastures and other browsing areas should be examined for bracken fern, which is native around the world, or other thiaminase-producing plants, and the plants removed or goats prohibited from those areas. This is particularly important for people living in an area where it is known to grow, especially if they have had a case of PEM in their herd and other causes have been ruled out.

From *Ruminations* #55

Treating PEM

4.5 mg per pound (for 200 mg/ml, that would be 1 ml per 45 lb), three times daily (first dose should be given IV), until symptoms of polioencephalomalacia disappear. In an emergency you can use B-Complex at about 12 ml per adult goat.

While the high doses of B-Complex will supply the needed thiamine (B1) it also hits the animal with an excess of the other B vitamins. Since the B vitamins are water soluble and quickly eliminated from the body via the kidneys there is little danger of an overdose, but, they do put an extra strain on the kidneys of an already stressed system. Straight thiamine (B1) is the best choice for treatment.

Note: Do not administer dextrose IVs to animals with PEM; their carbohydrate metabolism is impaired.

From *Ruminations* #28

Poisoning

Goats can be and have been poisoned by a variety of substances, including diesel fuel, insecticides, dewormers, nitrates and a variety of plants. Most of these poisonings can be avoided easily by ensuring that toxic substances are kept away from goats and stored separately from their food supply, as well. Plant poisonings can be prevented by surveying your pasture and other areas where the goats have access and destroying any poisonous plants there.

A common way for goats to be poisoned by plants is a neighbor or friend dumping cuttings from the garden or landscaping for the goats to eat. This has happened more than once with rhododendrons and azaleas.

A number of good resources, both in hard copy and online, list poisonous plants and their effects. The International Veterinary Information Service (www.ivis.org) has an online copy of the book *A Guide* to *Plant Poisoning of Animals in North America,* which is the best book on poison plants available.

Interestingly, goats sometimes eat poisonous plants and don't seem to have any reaction. This is because well-fed goats may eat only a small amount of a plant and then avoid it because they don't like the taste or have better things to eat. Greedy, dominant goats also may eat more of a poisonous plant and get poisoned, while the smaller, less aggressive ones avoid it. In such cases they don't get enough of the toxin to actually poison them. In other cases, food they ate previously may slow absorption of the toxin.

Different plants also may be toxic to one mammal but have no effect in others. Some parts of a plant may be poisonous, while the rest causes no reaction. In addition, the state of the plant may be the factor that causes poisoning. For example, in some plants wilted leaves concentrate the toxin, while fresh leaves may have a small enough amount to be safe.

According to an article by Jackie Nix, an animal nutritionist, "…common poisonous compounds found in plants include glycosides, alkaloids, oxalates, oils, minerals, resins and nitrates. Some of these poisons affect the nervous system, some the blood, and still others the intestinal tract or the heart. Knowledge of the specific poison and its mode of action will aid in trying to treat specific poisoning cases."

Symptoms of poisoning vary depending on the type of poison. They can include foaming at the mouth, vomiting, convulsions, staggering, crying, difficulty breathing or rapid breathing and heart rate, and in worst cases, death.

Because treatment may be different for different types of poisoning, the first step is to try to determine what may have poisoned the goat and eliminate that poison from the reach of other goats. If you have access to a veterinarian, call him or her.

Administer an activated charcoal product such as ToxiBan if the goat has been poisoned by an insecticide, rat poison, dewormer, organophosphate or alkaloid. This will absorb the toxins. If the goat has been poisoned by antifreeze, do not use ToxiBan. Another option, according to nutritionist Jackie Nix, is to administer charcoal tablets, fluids for hydration and mineral oil. This will absorb toxins, prevent dehydration and coat the intestinal lining.

If you need to induce vomiting, 2 tablespoons of salt on the back of the goat's tongue will work. (For rhododendron or other plant poisoning, see the recipes on p 203.)

Beware of Nitrate Poisoning

Nitrate poisoning is a noninfectious disease that can affect goats that have access to forage or water that has high levels of nitrates. The ultimate cause of death by nitrate poisoning is a lack of oxygen in the blood.

Normally, nitrates are converted to nitrites and then to ammonia during ruminant digestion. When fields become contaminated with animal waste and fertilizer run-off, the animals may be put at risk. Also of concern is accidental consumption of fertilizer by goats.

Some plants are more likely to accumulate toxic levels of nitrates. These include pigweed, oats, beets (not beet pulp pellets), lambsquarters, alfalfa, sudan grass, johnsongrass, and corn. Generally, the more mature the crop when it is cut, the less nitrate it contains.

As with many other conditions, healthy animals are less likely to be affected than those that are in poor health. In addition, consumption of carbohydrates speeds up the process of converting the nitrates, putting the goat at less risk.

If a goat shows signs of nitrate poisoning—dark blood, weakness, trembling, difficulty breathing and foaming at the mouth—a blood sample will be required to confirm the diagnosis. Nitrate levels may also be determined by testing feed or water, if that is suspected. The treatment of choice is intravenous methylene blue.

From "Health and Science News," *Ruminations* #36

Prussic Acid Poisoning
by Kelly Olsen

I headed to the barn to do the evening chores. The weather was hot, like always in Texas, but otherwise everything appeared to be quite normal. I usually feed grain first and then hay, but decided to do the opposite today since my boyfriend Laine was still cleaning inside the barn. I threw the goats three blocks of coastal hay to keep them occupied until he got to where he could stop. I went about the barnyard doing some of the other chores.

After 20 or 30 minutes I heard one of my junior does crying out as if she were in pain. I ran to the goats to see what was going on. She was lying down bawling. While I was trying to figure out what was wrong with her another one dropped to the ground crying out the same way, then another, until eight does were down. Then several of them started lying on their sides still bawling.

At first we thought maybe the heat was too much for them so we tried to give them water. That didn't help. They started thrashing around and having what appeared to be seizures, shaking, and breathing heavily and with difficulty.

As I looked around, I felt helpless. My girls were suffering and I had no idea why. I thought of the hay I had thrown, which had a small amount of johnsongrass mixed in it. Nathan (my son) and Laine removed all of the hay from the feeders. Laine had already gone for the phone to call our vet. I explained everything to her and she asked us to bring them in to the clinic. Laine, Nathan and I loaded eight goats into the back of his truck. (My stock trailer wasn't available for use.) Nathan stayed behind to keep watch over the rest of the herd. I rode in the back of the truck with two goats standing. The other six were lying down.

By the time we got to the clinic we had lost one doe. We carried all of the goats in. The vet couldn't find anything physically wrong other than the seizures. After the initial exam, she did blood work on the sickest-appearing and the healthiest-appearing doe. She found that the red blood cells were hemolyzing (breaking down). Something was inhibiting the cells in the goat's body from using the oxygen. They were suffocating.

We lost a second doe while she was on the examining table. At this point we still didn't know for sure whether the hay had caused the problem. The vet decided to do a necropsy on the second doe that died and send the contents to the lab for testing. The other six does, amazingly, survived. Five of them started recovering at the clinic. The sixth one still seemed out of it, but better than she was when we took her in. Two days later we took her back for more

Goat Health Care

blood work. She had developed an infection and was put on medication for it. She has recovered and now knows (unfortunately) what a needle looks like!

The test results revealed that the cause of the problems was prussic acid poisoning.

Prussic acid is a rapid-acting toxin that affects mammals. A number of common plants can accumulate large amounts of prussic acid, which is a cyogenic—cyanide-producing—compound. These plants include wide leaf grasses such as johnsongrass, other sorghums and sudan grass, and leaves such as those of apple, apricot, cherry and peach trees. More detailed information can be found at www.ext.nodak.edu.

Cyogenic compounds are located on the outer tissue of the plant, while the enzymes that enable production of cyanide are on the leaf tissue. The combination of these compounds with the enzymes creates prussic acid. Such a combination can occur when the plant cell is ruptured by cutting, wilting, freezing, drought, crushing, trampling, chewing or chopping. In my goats' case, the drought we underwent this year was determined to be the culprit.

Tip: Some of the plants that contain cyanide include catclaw, flax, apricot, peach, pincherry, chokecherry, pine cherry, wild red cherry, bird cherry, fire cherry, wild black cherry, elderberry, apple, Johnson grass, sudan grass, common sorghum, poison suckleya, white clover, velvet grass, arrow grass, goose grass, sour grass, pod grass and corn leaves.

Once eaten, prussic acid rapidly enters the bloodstream and inhibits oxygen, causing the animal to suffocate. Signs can occur as quickly as 15–20 minutes after consumption. These signs include excitement, rapid pulse and generalized muscle tremors, followed by rapid and labored breathing, staggering and collapse. A sign that hemolysis is occurring is bright cherry red, rather than pink, mucous membranes.

Thankfully, all of the survivors are happy and healthy again. At least until this drought ends and we get some normal rainfall I am buying my hay from a feed store that gets it from out of state.

I hope that by sharing this story of the horrifying experience my girls went through, other goat owners can learn and not have to deal with a similar situation.

From *Ruminations* #54

Q Fever

Q Fever is an infection caused by the bacterium *Coxiella burnetii*. Most goats with the infection do not show any signs, but can still pass it through manure, urine or birthing fluids. Humans can also get the disease from breathing contaminated barnyard dust, from drinking contaminated milk or from tick bites. Shedding of bacteria in the milk can occur for up to 52 days after birth. People with pre-existing heart valve disease or with valve replacements are cautioned to be extra careful around goats.

Q Fever is a worldwide disease, and outbreaks have been reported in Greece, the US, Canada, Australia and other countries. An estimated 20% of goats in Ontario, Canada and 26% in California have antibodies to *C. burnetii*. Goats are now considered the most common source of human infection.

The disease usually comes to light during kidding season, when it is spread to humans, or causes multiple abortions among goats in a herd. Abortions have been shown to occur 25–48 days after infection. Kids that do survive usually die within 24 hours of birth. *C. burnettii* is found in the mammary glands, lymph nodes above the mammary glands, placenta and uterus. It may be shed during birth and lactation. A 2003 study showed that the life of the infection is limited to two kidding seasons.

In at least one reported case, the presence of Q Fever in goats came to light when traced to infection in humans from goat manure used in the garden.

Diagnosis is by laboratory testing of the placenta. In the case of an abortion in a goat herd, the placenta should be refrigerated in the event that other goats abort and a cause needs to be determined.

Tetracycline is considered the drug of choice, and chloramphenicol is also effective. Pregnant animals should be segregated and placentas and dead kids should be burned or buried to help reduce spread of the disease. When purchasing new goats, inquire about whether the breeder has had any problem with abortion in the herd over the past several years.

Sources:

Hatchette, et al. 2000. Goat-Associated Q Fever: A New Disease in Newfoundland. www.cdc.gov/ncidod/eid/vol7no3/pdfs/hachette.pdf.

Hatchette, et al. 2003. Natural History of Q Fever in Goats. www.liebertonline.com/doi/abs/10.1089/153036603765627415?cookieSet=1&journalCode=vbz.

Wyoming State Veterinary Laboratory. 2004. Q Fever in Goats, Sheep, and People. http://wyovet.uwyo.edu/Diseases/2004/Qfever.pdf.

Rabies

While uncommon, from time to time a goat is diagnosed with rabies. In 1996 a goat with rabies was shown at a county fair in New York. Almost 500 people received treatment as a result of contact with that goat. In 2004, only six cases of rabies in goats were reported in the US; in Canada less than 1% of all rabies cases were in goats that year. Mexico reported only one case in 2004. In both 2007 and 2008, at least one goat with rabies was reported in the news.

When goat is found to have rabies, it must be euthanized and treatment is recommended for individuals who were exposed to the goat and may be at risk. Rabies is found throughout the US, except Hawaii, but is more prevalent in the eastern US. In some parts of the US, such as the county where the goat at a fair was found to have rabies, vaccination is required or recommended.

Rabies in a Goat: In 2008, a rabid goat was found on a farm in West Virginia. The goat was euthanized, and 20 other livestock were quarantined for eight months. Seven people who had contact with the goat were treated. A fox in the area had tested positive for rabies several months earlier.

Unfortunately, no rabies vaccine has been approved for goats, so it is considered an extralabel use and some veterinarians may refuse to use it. A Cornell University College of Veterinary Medicine program to study the use of a rabies vaccine in goats initially offered a possible solution, but it was discontinued in after a failure to raise the $75,000 needed to carry it out the first phase of the study.

What is rabies?

Rabies is caused by a virus that affects the central nervous system (CNS) of mammals, leading to progressive inflammation of the brain. The main indicator of rabies in goats is unusual behavior, which may lead to depression or partial paralysis of the back legs. Other signs are drooling and abnormal swallowing, which may at first appear as though the goat has something caught in its throat. In some cases, the classic form of rabies, which you may know from the movies, may occur in goats. The animal at first is nervous and agitated, but may cry continuously or become aggressive, attacking other animals and people. Ultimately, over a period of several hours to days, the goat will go down, have seizures and die.

How is rabies spread?

In the US, the virus is found primarily in wildlife and spread to goats and domestic animals by contact with saliva from a bite, scratch or other wound. Occasionally rabid wild animals, such as bats, skunks or raccoons, will enter barns, paddocks and other areas where they may come in contact with goats. The goats can be exposed when they investigate the new animal.

Unfortunately, the symptoms of rabies can vary, making early detection challenging. Because of this, and the fact that other diseases can look like rabies, goat owners should isolate animals that show suspicious behavior or other signs of rabies. This allows for both observation and avoidance of injury and spread to other goats or people. A veterinarian should always be involved in such a case.

How can rabies be prevented?

One of the best ways to prevent rabies in goats is to keep dogs and cats current on their vaccinations, so they are not susceptible and potentially a source of infection if bitten by a rabid wild animal. Another option, which may not be feasible, is to not have any cats or dogs.

Goat owners may choose to vaccinate their animals if they are likely to be exposed to potentially rabid animals and/or they are valuable to the owner.

Pastured animals are generally at greater risk although, as previously mentioned, they also may be at risk in a barn. The potential risk of exposure for goats is usually lower than annual vaccination costs. A veterinarian can help in making the decision about the need for vaccinating goats against rabies.

Sources:

MMWR Recommendations and Reports, May 23, 2008. Human Rabies Prevention—United States, 2008. *Recommendations of the Advisory Committee on Immunization Practices.* www.cdc.gov/mmwr/preview/mmwrhtml/rr5703a1.htm.

Ohio State University Extension Fact Sheet: *Veterinary Preventive Medicine, Rabies Prevention in Livestock.* http://ohioline.osu.edu/vme-fact/0001.html.

Personal correspondence with Mary Smith, DVM, Cornell University.

CDC, Public Veterinary Medicine: *Public Health. Rabies surveillance in the United States during 2004.* www.cdc.gov/rabies/docs/rabies_surveillance_us_2004.pdf.

Ringworm

Ringworm (dermatophytosis) is a common fungal disease, caused by a variety of fungi, which affects goats and other mammals, including humans. It can occur on any part of the goat's body, including horn, hoof or skin. It is generally considered to be a harmless disease.

It is transmitted through contact with infected animals, fomites (objects), such as fence posts, or other contaminants. The spores can remain in the environment for years. Cats can also spread it to goats, because they may have ringworm but show no signs.

Ringworm usually appears as a non-itchy, circular, scaly or crusty short-haired patch on the body. It often can be diagnosed by examination under a microscope of hairs from the area or a skin scraping.

In some cases the hairs taken for diagnosis are cultured in a special medium that inhibits bacterial growth and enhances fungal growth. Since a fungus is a slow-growing organism, diagnosis may take up to three weeks.

Ringworm should be treated to avoid spread to humans or other animals and to decrease infection of the environment. Small patches can be spot-treated with an antifungal cream, such as 1% clotrimazole cream. More widespread or severe infections may need to be treated with an oral medication, such as diflucan. Always wear gloves when treating ringworm.

In order to prevent ringworm in other goats in the herd, keep the infected animal isolated until the infection clears up. Thoroughly clean the isolation area before allowing other goats to use it. Try to keep cats, or other animals that are showing signs of ringworm, out of the goats' area.

Scrapie

Scrapie, a transmissible spongiform encephalopathy (TSE) is a fatal, degenerative central nervous system disease affecting sheep, and rarely, goats. It is related to "mad cow disease," or bovine spongiform encephalopathy (BSE), chronic wasting disease in deer and similar diseases in other animals.

Scrapie, as well as other TSEs, causes sponge-like spaces in the brain that ultimately lead to impaired functioning and death. Scrapie has been in the US for more than 70 years, and in Europe for more than 250. The cause of scrapie is unknown, and researchers even disagree on whether it is a protein called a prion or a virus.

It takes from two to five years after infection for a goat to show symptoms. It can be mistaken for a retrovirus, pregnancy toxemia, listeriosis, rabies or poisoning. It is believed to be spread from dams to kids through the placenta and amniotic fluid. It was recently found to be transmitted through milk and contact with sheep.

In Iceland scrapie recurred on farms 16 years after sheep were last there, causing speculation that it may stay in soil. One theory holds that a lack of copper, and a high level of manganese, may lead to the genesis of the disease. Interestingly, the areas of Iceland that are infected show very low levels of copper and higher levels of manganese. Perhaps that explains why scrapie is seen so rarely in goats.

Some sheep are genetically resistant to scrapie and a blood test has been developed to determine whether they are susceptible. Unfortunately, these genes have not been identified in goats. In 2007, an atypical form of scrapie was diagnosed in sheep that are genetically resistant to classic scrapie.

In 1998, a method was developed to diagnose scrapie with third eyelid tissue, rather than requiring necropsy. Ten years later a rectal biopsy was approved. In 2008 a blood test had not been developed, but research was underway.

As of 2009, only 21 goats had ever been diagnosed with scrapie in the US; in 2007 five goats in Michigan were found to have the disease, but the mode of transmission was unknown.

A national scrapie eradication program intends to eliminate scrapie from the US by 2010. It requires that sheep and goats be identified to allow tracing them to the farm of origin if they are ever found to be infected. A voluntary Scrapie Flock Certification Program (www.aphis.usda.gov/animal_health/animal_diseases/scrapie/) provides for testing and scrapie-free farm certification.

Soremouth

Soremouth, also known as orf (in humans), contagious ecythema (CE) and contagious pustular dermatitis (CPD), is a virus in the family Poxviridae. Known since at least the 1800s, the disease is found throughout the world where sheep and goats are located. While animals of all ages and breeds are susceptible, kids are more easily infected. Unlike most goat diseases, soremouth can also be transmitted to humans.

Course of Disease

Soremouth lesions begin to develop within 2–3 days of infection and within 11 days are visible. They usually start as blisters around the mouth, but can also affect the nose, the udder, the mouth, the throat, many other body parts and, in some cases, even the rumen and lungs. Young kids that get it can infect their mothers' udders, causing it to spread to other kids.

The sores last from a week to a month and can be spread to other animals or by objects that come into contact with the virus (known as fomites). The virus stays in the scabs and has been found in dried scabs for many years after the animal has recovered.

After the blister stage the lesions turn into pustules and become scabby when they break. Goats that contract soremouth usually develop a strong immunity and are not re-infected for at least a year.

While a malignant form of the disease has been reported, most animals recover easily as the disease runs its course. In cases where goats are exposed to thistles or other prickly bushes, they may have a harder time overcoming it, due to the skin breaks that let the virus in.

Soremouth that infects the inside of the mouth or the hard palate can be lethal. In addition, soremouth on a teat can lead to *staph aureus* and ultimately gangrene mastitis.

While soremouth is usually considered to be extremely infectious—with up to 100% of some herds showing signs—in other cases, only a few animals are affected. Young kids are more at risk for serious consequences. Where death occurs, it is generally as a result of secondary infection or certain fly infestations.

Treatment

Because soremouth is caused by a virus, it can't be treated with antibiotics. Infected animals should be kept clean, and the disease will clear up in one to four weeks with no treatment. When I had soremouth in my herd, I did an experiment, treating one group with tea tree oil, another with triple antibiotic and the other with nothing. All of them recovered at the same rate; apparently none of the treatments changed the course of the disease.

> **Tip:** Always wear gloves when dealing with what appears to be soremouth.

One source recommends mixing Vaseline and iodine to treat the lesions, and also to keep insects off of them.

Whatever treatment (or non-treatment) regimen is used, make sure to wear gloves and avoid exposure of any cuts or wounds to the active lesions, as the virus can spread to humans.

Prevention

A vaccine for soremouth is available, but it may cause the disease in herds that have not previously been exposed. It is a live vaccine, made by crushing up the scabs.

The best method for prevention is to either not bring new animals into your herd, or to quarantine those you do bring in. If a goat in the herd does show signs of soremouth, you can separate that animal until it recovers.

I tend to subscribe to the theory of exposing the other goats to mild cases of soremouth in order to develop immunity in the herd. I haven't had an outbreak since the initial one, using this method. But then maybe I'm just lucky.

Sunburn

Goats, particularly those that are white or have pink skin, can get sunburned after being clipped, or on very sunny days. White goats are also more likely to develop cancer, probably due to the lack of protection that pigment gives. Goat also an develop blisters or warts on the udder after spending too much time out in the sun.

> **Tip:** After clipping the udder or hair of a white or light-skinned goat, rub some baking soda on the clipped area to help protect against sunburn.

To prevent sunburn in such goats, limit their browsing time initally and then gradually increase it as time goes on. Others suggest using human sunscreen—but make sure to remove it prior to milking to avoid contaminating the milk.

Tetanus

Tetanus (lockjaw) is a well-known disease of goats and other animals. It usually comes about when a wound is infected by the bacterium *Clostridium tetani*, which lives in the soil. Tetanus can cause seizures and muscle rigidity. It is easily preventable with a vaccination.

One instance where tetanus can occur is disbudding; another is castration. Give a kid that is undergoing either of these procedures a shot of 150–250 units of tetanus antitoxin, which immediately protects against tetanus for several weeks. (Tetanus toxoid provides long-term protection against tetanus.)

Early signs of tetanus can be mistaken for other neurological disease, and include stiffness, difficulty opening the mouth and a "sawhorse" stance. As the infection progresses, the goat will become hypersensitive to touch and sound, and will ultimately collapse and go into seizures. Once a goat "goes down" with tetanus, it usually dies within 24 to 36 hours.

Dosage: For prevention of enterotoxemia and tetanus, inject 2 ml of *Clostridium Perfringens* C & D and Tetanus toxoid vaccine, SQ or IM.

Treatment for tetanus includes high doses of antibiotics and tetanus antitoxin, as well as hydration and treatment of the offending wound. Most goats with tetanus do not survive.

Prevention is easy and essential; all it really requires is vaccination of the animals, and cleanliness.

When an adult goat is known not to have been vaccinated, or its vaccination status is unknown, 500 to 750 units of antitoxin can be given when treating a wound or after a difficult kidding.

Preventing Urinary Calculi
by Cheryl K. Smith

Those who keep wethers as pets (and sometimes bucks) need to be aware that they can sometimes develop urinary calculi (stones). These are an accumulation of minerals or other compounds that can cause trauma to the urinary tract and obstruct the flow of urine out of the body.

One reason calculi are more common in wethers and bucks is because of their physiology—the urethra is a common site for blockage.

Sometimes medication or simple surgery can solve the problem, but unless the root causes are found, the animal may need to have further surgery or eventually be euthanized.

Types of Calculi

Urinary calculi can be of various types. If a goat develops and is treated by a vet for urinary calculi, insist that the stones be sent for analysis. Further treatment and prevention may depend upon what caused the problem.

Three major types of calculi can cause problems in goats: phosphate, silica and calcium. Each of these may be caused by feed rations.

Phosphatic Calculi. Goats that eat rations high in phosphorus, such as cereal grains, can develop struvite (magnesium ammonium phosphate) calculi. Balancing the dietary calcium:phosphorus ratio is essential. Ideally it should be 2:1. Feed tags for the rations fed to goats list percentage of contents, so that is a good place to start.

Ruminant saliva is rich in phosphorus and the route of excretion for phosphorus is the gastrointestinal tract. When the goats are fed pelleted grain, they produce smaller amounts of saliva, which then decreases the gut excretion of phosphorus, sending more to the urinary tract.

Silica Calculi. This type of urinary stone mainly affects sheep and cattle that graze the native grasses of western North America. As the grasses mature, the silica content tends to increase. Since the silica is not broken down in the rumen or bloodstream, if the animals don't drink enough water, they can develop silica calculi.

Calcium-based Calculi. Calcium carbonate and calcium oxalate urinary stones are found most commonly in sheep grazing lush, fast-growing clover pastures. This is because they are loaded with calcium, but have a low phosphorus and high oxalate content. The combination of oxalate and calcium makes the calcium unavailable for absorption. This makes the urine alkaline and leads to the development of calcium-based stones. In North America calcium-based stones have been found in goats that ate mainly alfalfa.

Castration

Wethers can be more prone to urinary calculi because they are castrated. Since castration at an early age has been shown to be a risk factor for development of urethral obstruction, avoid wethers that are castrated at less than three months, if possible.

A study done in cattle showed that although calculi can form in the urinary tracts of both bulls and steers, bulls may be able to pass a stone that would be likely to obstruct the urethra of a steer. This is because the testosterone produced by the bull makes the diameter of the urethra 25% bigger than that of a steer. This is probably also true in bucks and wethers for the same reason.

Water Consumption

The most important factor in preventing urinary calculi is to increase the water consumption of the goat. Keep water bowls clean and fill them with fresh water frequently. If you use automatic waterers, use shallow containers that can be refilled rapidly.

Use heaters or plug-in buckets during the winter or make sure to give your goats hot water regularly to encourage more consumption. During the summer, make sure that the water is in a shady spot.

In a large herd, use multiple watering sites so that all goats, regardless of their position in the herd, have access. Some people have found that flavoring the water with sugar-free drink mixes will increase the water intake of their animals. According to Canadian studies, the mineral content or hardness of water does not play a significant role in causing stones.

Feed Ration

Wethers and bucks should be fed grass hay as their main source of forage, to prevent the development of calcium-based stones. Alfalfa and other types of legume hays contain more calcium than is healthy for them. Be aware that avoiding these types of hay does not guarantee the prevention of urinary stones; a genetic influence may still lead to problems.

Wethers and bucks should also not be fed large quantities of grain; as noted above, this can lead to phosphate stones. Whatever you decide to feed, remember to balance the calcium and phosphorus.

Frequency of Feeding

Because of chemical reactions that take place affecting urine concentration at each feeding, frequent or free choice feeding is considered to be another factor in limiting urinary stones. Concentrated urine can cause urinary stones. Rather than feeding the goat once or twice a day, teach free choice feeding from a young age, or give small portions several times a day.

Salt in the Diet

Increasing the salt concentration in the diet promotes more water intake, leading to more diluted urine. According to studies in Canada, loose or lick salt provided free choice was found not to be adequate in preventing urinary stones in animals that were considered to be at risk for silica calculosis.

They found that mixing the salt directly into the feed was the most effective means of providing it to the animals. Some people give corn chips with extra salt as a treat; others add extra salt to moistened feed. Spraying a salt solution onto hay, for those that do not feed grain, also can be helpful. The recommended amount of salt in the diet is 3–5% of daily dry matter intake.

Ammonium Chloride

Ammonium chloride is another option for preventing urinary calculi. It should be fed at a level of up to 1% of dry matter in the diet. The action of ammonium chloride is to reduce the pH of the urine (make it more acidic), which will make various types of urinary stones more soluble in the urine. Molasses should be avoided as a way to get the goat to drink more water or to make the ammonium choride more palatable, since its high potassium may make the ammonium chloride less effective. Sugar is a more effective alternative.

The downside of feeding a wether ammonium chloride over a long period of time is that it has been shown to reduce the mineral content in the bones in ewes. It may have the same adverse effect on a goat.

Conclusion

Urinary calculi affect only some goats. Following these guidelines with all wethers is a good way to prevent the problem before it even gets started. If a wether does develop a problem, you can still take these necessary steps to prevent a recurrence in the animal and possibly save its life.

Source:

Van Metre, David C. *Urolithiasis in Small Ruminants: Surgical and Dietary Management.* Colorado State University.

From *Ruminations* #51

Medications

Extralabel Drug Use

Because goats are considered a minor species, many of the drug products developed and available for use are not approved by the Food and Drug Administration (FDA) for use in goats. In addition, sometimes the dosage needed for goats differs from that found on the bottle, both in volume and in length of time to be used.

Using a product that is not approved and labeled for use in goats constitutes "extralabel" drug usage. Much of the problem is due to the relatively small numbers of goats in the US—a problem that may be resolved with the growing marketability of meat goats. In the past, drug companies have not been able economically to justify going through the testing required to approve a drug for use in goats. Simple experience, coupled with information derived from veterinarians who specialize in goats determines effective doses in the caprine.

The fact that a product is unapproved for use in goats does not indicate that it is unsafe or ineffective.

> Extralabel use is "use in a manner that is not in accordance with approved labeling." Such use requires that:
> - a valid veterinarian-client-patient relationship exists;
> - no approved animal drug labeled for the treatment need has the same active ingredient in the required dosage form and concentration; or, the approved animal drug is clinically ineffective for its approved use and an effective substitute is needed.
> - the veterinarian carefully evaluate and diagnose the condition requiring treatment.
> - the veterinarian establish a scientifically appropriate withdrawal period, based on appropriate scientific information, if available.
> - The veterinarian ensure that the treated animal's identity is carefully documented and maintained.
> - the veterinarian ensure that the assigned withdrawal times are observed and no illegal drug residues occur in a food-producing animal receiving extralabel drug treatment.

Doses and Important Medication Notes

A dose of a drug "for mature does, bucks, kids" refers to the large goat breeds, not to Nigerian Dwarves or other mini dairy goats.

For some drugs and most vaccines, "one size fits all," but for others dosing by weight is essential. *Please read the suggested dose material to determine the correct dose for the smaller breeds.*

ALWAYS READ LABELS to make sure to use the proper medication and appropriate strength, and to avoid over- or under-dosing (many drugs come in more than one strength), as well as to obtain additional dosing information, precautions, indications, withdrawal times, etc. If using the product extralabel, write those instructions on the bottle.

Also see extralabel information above. Withdrawal and milk discard times are for FDA-approved dose rates and routes of administration, and unless otherwise noted are for cattle. Extralabel doses may result in longer withdrawal and milk discard times.

For FDA-approved drug information, visit and register with FARAD (Food Animal Residue Avoidance Database, www.farad.org/), and the FDA Center for Veterinary Medicine (www.fda.gov/cvm/).

DO NOT MIX MEDICATIONS IN THE SAME SYRINGE OR WITH ANOTHER PREPARATION without checking compatibility of medications. Drug incompatibilities can change the chemical or physical nature of drugs. Incompatibilities can occur between two drugs or between a drug and carrier, the environment or even the receptacle (IV tube, etc.).

READ AND FOLLOW MILK AND MEAT WITHDRAWAL REQUIREMENTS. The intent of determining these times is to keep medications out of the food chain. Do not drink milk within the withdrawal time and don't sell a goat (for example, at a livestock auction) until the meat withdrawal time has passed.

Unstable medications usually have a short shelf life when in solution. Always label reconstituted solutions with the new expiration date. If indicated on the label, refrigeration or freezing can prolong the shelf life. However, *do not* assume this is true of all medications: READ THE LABEL. Freezing can increase the degradation of some medications (e.g., ampicillin). Refreezing a previously frozen solution increases the risk that it will not be as effective.

> US Food and Drug Administration (FDA) policy on residues "is to hold responsible any individual in the production and marketing chain who can be shown to have been responsible for having 'caused' (by any act of commission or omission) illegal drug residues in edible animal products."

When giving antibiotics, always give a full course, using an adequate dose at the recommended frequency to minimize the development of resistance. Continue even if symptoms have disappeared! The hardier bacteria are still present and must be killed or stopped from reproducing. Stopping antibiotics early often causes a recurrence, or allows the stronger bacteria to develop a resistance to the antibiotic. Continue to treat for 24–36 hours after asymptomatic (without symptoms).

Important Accessories and Equipment

Alcohol Wipes: Pre-packaged alcohol wipes can be found in drug stores or through medical supply companies. They are good for sterilizing thermometers, medicine bottle tops, equipment or injection sites.

Catheter: A must for tubing a weak or premature kid at birth. You can buy catheter and syringe set-ups (60 ml syringe and catheter) from most catalog houses for just a few dollars.

Cotton Balls: Keep in a sealed container (small Rubbermaid or Tupperware containers work great) filled with 70% ethyl alcohol. For cleaning injection sites, thermometers, medicine bottle tops, equipment, etc.

Feeding Syringe or Drenching Gun: For administering herbal or chemical dewormers, other medicines or water to a sick or weak goat.

Film Canisters: These are useful for filling with iodine for dipping newborn cords.

Gloves: Disposable latex gloves can be found at drug stores, medical supplies stores or through veterinary and pet supply catalogs.

Needles: 20 X ¾" or 20 x 1" (most useful, all purpose). Can take thicker medications. 22 X ½ " or 22 X ¾" are less painful; they can be uses for babies and thin liquids. (Optional, but handy.) 18 X 1" for very thick medications, can be used for bucks and larger does, and for painful injections where rapid injection is useful.

Scalpel with disposable blades: For lancing abscesses or other minor surgical procedures.

Scissors, surgical: For cutting umbilical cords, bandages, etc.

Sharps Container: Used for proper disposal of needles. See section on Biosecurity for more information.

Stethoscope: To listen to lung sounds or rumen sounds, or fetal heartbeats in pregnant does. While not necessary, it can come in handy, and with practice you will learn to "hear." Listen to lung and rumen sounds on healthy animals and you will be able to hear unhealthy sounds when present, e.g., the lungs of a goat with pneumonia often sound like an ocean (fluid filled) or raspy, like sand paper rubbing together. Detecting pregnancy by hearing fetal heartbeats is fun.

Stomach Tube: A foal-sized stomach tube is handy for tubing larger goats. You can use any small diameter 1/4"–5/16" hose or tubing, but a standard tube is made of good quality PVC material that is soft and flexible, will last a very long time and has an attached funnel end. Measure the distance from chin to mid abdomen, the same as for tubing kids.

Syringes: 3 ml, 6 ml, 12 ml, 20 ml, 30 ml, 60 ml. (3 ml and 6 ml most frequently used.) The Luer Lock syringes are preferable, as they are more likely to stay secured and not pull off. You can buy syringes with a "feeding tip," a larger tapered tip that is used for dosing liquids and or tubing weak newborns (also, see catheter). You can also use 30 ml or 60 ml feeding tip syringes for giving oral medications.

Thermometer: This is the *#1 PIECE OF EQUIPMENT* a goat owner can have; it is a window to the inside of your goat. Use whenever an animal is "off" for any reason (temperatures usually rise 24–36 hours before other signs of illness). It will tell you if your treatment is working or not; an elevated temperature should start to drop in approximately 24–36 hours after beginning antibiotics if the medication(s) is combatting the infectious bacteria. If the temperature has not gone down after that time you should consider changing to another medication. (Note: Do not count temperatures reduced by other medications, e.g., Banamine, Dipyrone, aspirin, bute, etc.)

Anaphylactic Reaction

Allergies are hypersensitivity reactions of the immune system to specific substances (allergens, such as pollen, stings, drugs or foods). The most severe form of allergy is Anaphylactic Shock, which is a true emergency. Anaphylactic Shock is classified as a Type I Hypersensitivity (IgE-mediated Hypersensitivity, Immediate Hypersensitivity). Although the technical term "anaphylaxis" means this type of immune reaction, even if mild, most doctors, veterinarians and lay people use it to mean a life-threatening rapid allergic reaction.

In animals it most often follows an injection, often a repeat injection of a substance. The animal collapses within seconds or minutes afterward. Epinephrine (also known as adrenaline) must be administered immediately.

Never give shots of any type without having epinephrine on hand! (Note: the only anaphylactic reaction one of my animals ever had was to a vitamin A, D & E injection.)

Drenching

To give a goat medication with a feeding syringe or drenching gun:

1. Restrain the goat by having someone else hold it and back it into the corner. Hold the upper neck with your left arm (or right arm, if you are left-handed.) Hold the drencher in your other hand.

2. Insert your thumb into the goat's mouth and pry it open, wide enough to insert the neck of the drencher. Don't put your fingers into the mouth or you will be bitten by the grinding teeth in the back.

3. Insert the neck of the drenching gun into the corner of the goat's mouth about 1-½ or 2 inches, while angling the goat's head to about 45 degrees so the medication will go down its throat.

4. Slowly press the plunger to deliver the medication to the rear of the tongue and allow the goat to swallow it. Don't go too fast; the liquid could go into the lungs and cause aspiration pneumonia.

Biosecurity

Properly dispose of used needles, syringes and other used "medical" objects. Sharps and other piercing objects can puncture regular waste bags, posing both a physical and contamination hazard.

What is medical waste? Hypodermic needles and any items that have come in contact with human or animal specimens, culture media, DNA, live and attenuated vaccines, including:

- Needles, razors, and scalpel blades
- Broken glass and plastic
- Pipettes, pipette tips, capillary tubes
- Sharp-cornered objects

Acceptable Containers. These containers must be sturdy, leakproof and puncture-resistant. Generally they are sturdy, lined cardboard boxes or plastic jars. Most hospitals will accept and dispose of your filled containers for you, as will some local fire departments. Other possibilities are veterinarian, MD, dentist, etc. You can purchase and use a regular Sharps Container; milk/bleach bottles, etc., work well, too.

Adapted from *Ruminations* #27–29

Routes of Administration

PO—per os, by mouth
SQ—subcutaneous, under the skin
IM—intramuscular, into the muscle
IV—intravenous, into the vein

Dosage Conversion Factors
To convert mg/lb to mg/kg:
mg/kg = mg/lb X 2.2
To convert mg/kg to mg/lb:
mg/lb = mg/kg X 0.454

Practitioners experiencing drug-related problems should contact the drug's manufacturer and the Food and Drug Administration's (FDA) Adverse Reaction Hotline—301-594-1722—during working hours of Monday to Friday, 7:30–4:30 EST; 301-594-0797 after hours; FAX 301-594-1812, so that the incident can be documented. The FDA's Adverse Drug Reaction (ADR) information is published annually in the FDA Veterinarian and is also available through the FDA's home page at www.cvm.fda.gov under the Office of Surveillance and Compliance category.

Injectable Antibiotics

Recheck ALL withdrawal times for medications below with your veterinarian. Extend withdrawal two to three times for ALL medications that are used in a manner other than labeled and for animals undergoing extended treatments.

Oxytetracycline (Extralabel, OTC)

LA-200®
Oxy-Tet 200®
Bio-Mycin 200®
Liquamycin 200®
Geomycin 200

- All contain 200 mg of oxytetracycline per ml.
- All are labeled for SQ injection.
- *Tylan 200 is not the same medication.*

Withdrawal, Meat: 29 days. **Milk:** 6 days.

Storage/expiration: As they age, tetracyclines tend to be degraded and become more toxic.

Oxytetracycline is broad spectrum, effective against a variety of bacteria both gram positive and gram negative, and infections caused by rickettsiae, some *Mycoplasma*s and *Chlamydia*. Many *staph* (an opportunistic gram-positive bacteria) and strep organisms show increasing resistance to oxytetracycline. Most strains of *E. coli, Klebsiella, Enterobacter* and *Pseudomonas aeruginosa* are also resistant to oxytetracycline.

Oxytetracycline Dosage:
5 ml/100 lb (9 mg/lb) SQ every 48 hours for 4 days.

Often combined with sulfas (e.g., Albon [sulfadimethoxine]) with which they are synergistic.

A good therapy for metritis (uterine infection) following freshening, navel or joint ill in young kids (always treat joint/naval ill for a full 10 days), and is the preferred medication for the most common causes of contagious keratoconjunctivitis ("pinkeye") in goats, *mycoplasma* and *Chlamydia*. Also used for treating foot rot, pneumonia and abortion outbreaks.

LA-200 injections can be painful, Oxy-Tet 200 and Bio-Mycin 200 use a different "carrier" that is not painful or as irritating to tissues when given SQ and rubbed in well.

In addition to being injected subcutaneously, oxytetracycline can also be combined with sterile saline solution (5 ml LA-200, Oxy-Tet, or Bio-Mycin 200 with 15 ml sterile water) and infused directly into the uterus.

Also effective against most mastitis-causing bacteria via systemic injection.

Cautions: In young animals high doses or chronic administration may delay bone growth and healing. It should be avoided during early pregnancy if possible. Can cause photosensitivity (sensitivity to the sun).

Interactions: "Natural medicine" (herbs) and oxytetracycline: According to some reports, goldenseal, barberry and grape seed extract may reduce the effectiveness of antibiotics, specifically tetracycline and tetracycline-derivatives.

Procaine Penicillin G (Extralabel, OTC)
Various brand names

300,000 units Procaine Penicillin G/ml

Use 10,000 units/lb twice daily, with higher doses for well-established infections.

Withdrawal, Meat: 16–21 days. **Milk:** 5 days.

Note: Penicillin G is distributed into milk; in food animals the distribution is sufficient to cause residues in violation of law. However, (label dose) the concentrations of penicillin produced in milk are subtherapeutic for most bacteria. In sheep, 0.11% of an intramuscular injection of sodium penicillin G was distributed into the milk.

Procaine Penicillin Dosage:
Kids: 1–2 ml SQ for baby goats (8–25 lb), 1 or 2 times daily.

Adults: 3–6 ml/100 lb SQ for adult goats, 1 time, or double recommended maximum dose on first injection, then divide daily dose and give every 12 hours for 2–3 days, and once daily thereafter.

Used for treating gram positive infections, particularly streptococcus.

There is widespread resistance to penicillin.

Penicillin is often administered along with Naxcel. If you use one of these combinations, administer via separate syringes, do *not* mix together in the same syringe.

Blood levels tend to drop starting 12 hours after injection. Goats may be dosed every 24 hours, if you are unable to treat more often or at the end of an illness, once it is well under control.

Penicillin, Long Acting (LA) (Extralabel, OTC)

Penicillin procaine/penicillin benzathine combo

There is widespread resistance to both penicillin procaine and benzathine, and studies have shown that the benzathine (long-acting) form of penicillin has too slow an absorption time to reach recommended therapeutic levels.

(Note: Penicillin G benzathine and penicillin G procaine combination has been replaced by other more effective medications.)

Ceftiofur Sodium (FDA-approved for limited use in goats, Rx)

Naxcel®
Excenel®

Approved for lactating goats **only** for the treatment of bacterial pneumonia due to *Mannheimia* (formerly *Pasteurella*) *haemolytica* and *P. multocida* in the US.

Withdrawal: Zero days withdrawal when used at label dose.

Often used in conjunction with penicillin.

Naxcel has a very short shelf life (one week) once the powder is reconstituted, but it can be frozen for up to eight weeks.

Excenel is the same ceftiofur antibiotic as Naxcel, in a different base. *Advantages are a longer shelf life (Mfg. recommendation: 14 days after the first dose is removed), and no refrigeration required.*

> **Naxcel Dosage:** 1 ml/50 lb IM or SQ once a day, for 3–5 days.

Drug Family: 4th generation Cephalosporin.

Many goat people believe that Naxcel is the drug of choice for mastitis, although most people in the dairy cow industry disagree, noting that concentrations in the udder are too low to be effective.

Also useful for foot rot and pneumonia.

Florfenicol (Extralabel, Rx)

Nuflor®

300 mg/ml of florfenicol

Approved for beef cattle and non lactating dairy cattle.

Withdrawal, Meat: 38 days following one SQ injection. 28 days for two-dose IM treatment.

Milk: The FDA has not established a milk discard time, tolerance or safe level for florfenicol in milk.

> **Nuflor Dosage:** 6 ml/100 lb SQ one time, or 3 ml/100 lb IM in the neck area every 24 hours.
>
> Also effective is 6 ml/100 lb SQ every 4 days, three times.

Studies in goats indicate Florfenicol concentrations in milk are equal to serum concentrations.

Florenicol has a broad spectrum of activity against both gram-negative and gram-positive bacteria and is primarily bacteriostatic. Because of its low protein binding and extensive tissue distribution, florfenicol reaches higher levels in tissue than serum, reaching clinically effective concentrations at sites of infection. Studies on pharmacokinetics have also been conducted in horses, goats and pigs with Nuflor, although there are no label recommendations for these species.

Adverse reactions: Possible temporary decrease in feed and water intake. Diarrhea has been reported in cattle. Local tissue reaction and soreness at IM injection site. This injection appears to be very painful to goats.

> **Tip:** Nuflor is oil-based and goes in slowly; use 18 or 20 gauge needle to avoid extended syringe filling and injection time.

Note: Florfenicol belongs to the same antibiotic family as chloramphenicol. Florfenicol differs chemically and is not linked to chloramphenicol's human toxicity concerns (bone marrow suppression and aplastic anemia in humans). The FDA prohibited the use of chloramphenicol in all food-producing animals in 1984.

Tylosin (Extralabel, OTC)

Tylan 200® — 200 mg tylosin per ml

Note: Tylan also comes as Tylan 50 (50 mg tylosin/ml); check bottle for strength and adjust the dosage.

Tylan 200 is NOT the same as LA-200.

Tylan 200 Dosage: 1.5–2.0 ml SQ for baby goats, 3.5–5.0 ml per 100 pounds (7–10 mg/lb) IM or SQ for adults, one time daily. Use for no more than 5 days, or 24 hours after symptoms disappear.

Intramuscular bioavailability — Goats: 72.6% (15 mg per kg of body weight mg/kg] dose)

Withdrawal, Meat: 30 days. **Milk:** 96 hours.

Not approved for lactating animals in the US. Outside the US it has a 96-hour withholding/withdrawal time in lactating cattle, and 8–21 days in beef cattle. Tylosin concentrations in milk can be much higher than concentrations in serum.

> **Tylan 200 Dosage:**
> **Baby Goats:** 1.5–2.0 ml SQ for baby goats.
> **Adults:** 3.5–5.0 ml per 100 pounds (7–10 mg/lb) IM or SQ, one time daily. Use for no more than 5 days, or 24 hours after symptoms disappear.

Adverse reactions: Tylan injections are painful and the animal may develop a painful swelling at the injection site that may last for some time, particularly with an IM injection.

A good antimicrobial for upper respiratory infections and some forms of enteritis. The only antibiotic that really hits mycoplasma arthritis in young kids (see below).

Drug family: Macrolides

Mycoplasmal Polyarthritis, which affects kids at 4–6 weeks of age, will usually only respond to massive doses of Tylan 200 [40 mg per lb = 4 ml per 20 lb) 3 X daily for 2 days, then 2 X daily for 3 days and 30 mg per lb (3 ml/20 lb) 1 X daily for 10 additional days.

Mycoplasma must be treated for a 12–14 days.

All printed veterinary references state that while this treatment will get the active disease under control, but the "cured" animal will often be a carrier for life and will be able to pass it on to its offspring through the colostral milk. Anecdotal experience with mycoplasma (*mycoplasma mycoides ssp. mycoides* [large colony]) indicates that it does not return in an animal if full course of treatment is given and there is no evidence that it is passed to offspring.

Albon and Di-Methox 40% injectable (Extralabel, Rx)

Sulfadimethoxine 400 mg/ml (40%)

Broad spectrum, Coccidiostat.

> **Di-Methox Injectable Dosage:** 1 ml/16 lb IV or SQ initially; reduce by half to 2–5 days.

Erythromycin (Extralabel, Rx)

Gallimycin®

Withdrawal, Meat: 5 days. **Milk:** 96 hours.

Erythromycin is an old antibiotic, a "re-discovered" favorite in human medicine now, also sometimes used by dairy cattle producers to treat strep mastitis.

Bacteriostatic at low concentrations, bactericidal at high concentrations.

> **Gallimycin Dosage:** 1 mg/lb SQ once a day.

Effective against gram positive bacteria + *Mycoplasma, Chlamydia, Rickettsia*. (Note: There is some indication that intramammary infusion of erythromycin may cause allergic reactions in some goats).

Amoxicillin

Amoxi-inject® (Extralabel, Rx)

Dosage: 5 mg/lb SQ once a day.

> **Amoxicillin or Ampicillin Dosage:** 5 mg/lb SQ once a day.

Withdrawal, Meat: 26 days. **Milk:** 120 hours.

Ampicillin (Extralabel, Rx)
Polyflex®
5 mg/lb SQ, once a day.

Withdrawal, Meat: 10 days. **Milk:** 72 hours.

Enrofloxacin (**Prohibited in goats. Do not use.**)
Baytril® 100
Used in bovine respiratory disease (BVD)

Gentamicin (**Do Not Use**)
Gentocin®

Tilmicosin (**Do not use**)
Micotil®
Micotil (tilmicosin) can cause fatal reactions in goats (particularly toxic to Alpines and Alpine crosses).

Injectable Vitamins and Minerals

Vitamins A & D (OTC)
Check label for strength/doses.

Vitamin A helps release vitamin D, which makes calcium more available in the system.

Vitamin B-Complex (OTC)
Check label for strength/doses.

Dosage: 5–6 ml SQ (adult dose) for debilitated animals, those with enteritis, and those that won't eat. It helps to soothe the intestinal linings, and stimulates the appetite.

Vitamin B-12 (Rx)
Comes in different strengths; the most common are 1000 mcg/ml and 3000 mcg/ml.

Dosage: 1000–2000 mcg SQ for mature animals.

B-12 will restore appetite in a debilitated animal. I often boost B-complex with additional 500–1000 mcg of B-12.

Thiamine (Rx)
B vitamin (B1) that usually comes in 200 mg/ml strength.

Use SQ/IM to treat polioencephalomalacia (goat polio).

> **Thiamine Dosage:** For goat polio, give 1 cc/35 lb up to three times per day IM, SQ, or orally.

Vitamin E (OTC)
Check label(s) for strength and dosage.

> **Trick:** When treating mastitis, supplement vitamin E with an injection or oral dose of 3 ml Bo-Se, if it hasn't been administered within the last 2–3 months.

Vitamin E works in conjunction with selenium and is essential for tissue health (including udder health). Human vitamin E capsules daily in grain or injectable form (Vital E, etc.) every two to three months, or when ill or severely stressed.

Selenium Tocopherol (Extralabel, Rx)

The most commonly selenium formulations used in goats are Bo-Se® or Myosel-B®. Both contain 1 mg of selenium and 68 IU of vitamin E per ml.

Dosage: Bo-Se/Myosel-B — 1 ml/40 lb; Mu-Se/Myosel-M, Velenium — 1 ml/200 lb. Some veterinarians recommend up to 3 ml of Bo-Se several times a year.

Withdrawal, Meat: 30 days for beef cattle in the US.

Caution: Brand names and strengths vary. Mu-Se (Myosel-M & Velenium®) contains five times as much selenium as Bo-Se or Myosel-B. ALWAYS check product NAME and label for doses. *Avoid giving it with any other vaccines or drugs.*

Note: Make sure that the drug date is not expired, as the vitamin E goes bad and it may be ineffective.

Tip: When injectable selenium is not available, give a kid that is exhibiting symptoms of white muscle disease two tablets of 200 mcg selenium at once. Do *not* repeat this dose immediately, as any selenium not used will be stored in the liver. If necessary, repeat in two weeks.

You can alternatively give 1000 units of oral vitamin E at birth, as a one-time dose.

White muscle disease is caused by a deficiency of selenium, and is occasionally seen in young stock. Bo-Se/Myosel-B injections are routinely given to both bucks and does at about one month of age, to bucks in the early fall before breeding season begins, and to pregnant does 30–45 days prior to freshening.

Particularly helpful in the western US, and some other areas, due to selenium deficiency in the soil.

A deficiency in selenium can lead to lowered resistance to disease, reproductive problems, and a variety of other deficiency signs. Recent research points to a definite link between udder health, selenium and vitamin E.

Be aware of whether your part of the country has high or low selenium soil concentration. In areas with a high concentration, goats may develop selenium toxicity if given selenium. Symptoms include staggering, muscle weakness, abortions and kidney failure. Talk to the county extension agent about selenium in your area and whether you should supplement it.

Miscellaneous Injectables

Epinephrine

For treatment of anaphylaxis and in cardiac resuscitation.

Dosage: Ruminants: 0.5–1.0 ml/100 lb body weight of the 1:1000 SQ or IM: DILUTE to 1:10,000 if using IV; may be repeated at 15 minute intervals.

Tip: Never give an injection of any type without having epinephrine on hand!

Note: Be certain when preparing injection that you do not confuse 1:10 (1 mg/ml) with 1:10,000 (0.1 mg/ml) concentrations. To convert a 1:1000 solution to a 1:10,000 solution for IV or intratracheal use, dilute each ml with 9 ml of normal saline for injection.

Store epinephrine in a tight container and protect from light. It will darken (oxidation) upon exposure to light and air. Do not use if it is pink, brown, or contains a precipitate.

ANAPHYLACTIC REACTION (SEVERE, IMMEDIATE ALLERGIC OR SHOCK REACTION). The animal suddenly collapses after a shot or shock situation. Epinephrine (sometimes known as adrenaline) must be administered immediately.

Flunixin Meglumine (Extralabel, Rx)

Banamine®
Citatron®

Both contain flunixin meglumine, 50 mg/ml.

Flunixin meglumine is a potent nonnarcotic, nonsteroidal, analgesic agent with anti-inflammatory and anti-fever activity.

Banamine Dosage: 1 ml/100 lb IM, SQ or IV given as a single dose every 24 hours for 3–5 days.

May dilute with vitamin B-12 injection to reduce pain on IM injection. (Some vets report abscess at the injection site and recommend cleaning the area with alcohol first.)

Banamine also comes in paste and granules.

Contraindications: Gastric ulcers, concurrent kidney disease, untreated dehydration, other NSAID use. Limit treatment to once daily for as few days as possible. Long-term use not recommended (possible ulcers, kidney problems).

Withdrawal, Meat: 14 days. **Milk:** 4 days.

In the US, Banamine® brand of flunixin meglumine only was approved (mid '98) for non-lactating and beef cattle.

Note: Banamine can be used to reduce fever, increase white blood cell motility, stimulate interferon production and activate T cells. It also inhibits the growth of many microorganisms. *A low to moderate temperature in an animal that is "off," but still feeling well enough to eat and drink, etc., sometimes is best left untreated.*

The effects of Banamine last 12–24 hours. While normally prescribed for use only once every 24 hours, it is sometimes used at 12-hour intervals for acute problems.

It has some anti-endotoxic activity when used as part of treatment for toxin producing infections (mycoplasmosis, pasteurellosis, etc.). It also is used in respiratory infections to combat attendant inflammation of the lungs (relieves coughing and dyspnea [difficulty breathing] and areas of consolidation in the lungs). It helps to control respiratory inflammation so the animal feels and eats better.

Banamine is expensive, but well worth using in a debilitated animal, as it allows for fast reduction of symptoms and thus return to eating, which helps greatly to speed up recuperation while other medical therapy is taking effect (e.g., antibiotics).

Toxicity: ALL NSAIDs work by blocking prostaglandins (see above), chemicals that cause the symptoms of pain and inflammation. Unfortunately, "good" prostaglandins maintain blood flow to the kidneys and the lining of the stomach and intestines.

Banamine and other NSAIDs may cause kidney damage and ulcers in the stomach and intestines. The risk of toxicity is greater in the very young, and in old animals.

Clinical signs of toxicity include teeth grinding and drooling, low grade colic pain, diarrhea and fluid accumulation on the abdomen and legs. Kidney failure and perforation of stomach and intestine ulcers can be deadly.

Banamine Is Effective Given Orally

A study done using six Norwegian dairy goats looked at the administration of flunixin (Banamine) either intravenously, intramuscularly or orally. Flunixin is a nonsteroidal anti-inflammatory drug (NSAID) used for pain in goats.

The dose given in the study was 2.2 mg/kg of body weight. The researchers found that the blood levels and length of time it stayed in the body were comparable, regardless of method of administration. It was found to stay in the body for an average of 22 hours. This means that for those goat owners who are uncomfortable with giving injections, but need or want to treat their own goats, giving Banamine orally will work just as well.

From "Health and Science," *Ruminations* #56

Dexamethasone (Extralabel, Rx)
Azium®

Most commonly 2 mg/ml of dexamethasone.

Caution: Do not use in pregnant does, it can cause them to abort.

Dosages: As an anti-inflammatory and pain reliever for joint and bone injuries, use 0.5–1.0 ml/20 lb.

For head injuries or "brain burn" following too vigorous disbudding, use 1–2 mg/20 lb.

For shock, give 1–2 mg/20 lb. This will 1.) increase capillary blood flow (improve circulation), 2.) decrease absorption of endotoxins, 3.) decrease production of Myocardial Depressant Factor, and 4.) decrease organ damage.

Following stroke or other cerebral vascular accidents, use 1–2 mg/20 lb.

For ketosis, give 4–8 ml.

With allergic reactions to insect bites or other allergens, give 0.5–1.0 ml/20 lb. (The anti-inflammatory effect of 0.75 mg of dexamethasone approximately corresponds to 5 mg of prednisolone or 20 mg of hydrocortisone.)

When used as supportive therapy while an animal is recuperating from severe debilitation (and therefore eats better during the very critical period of early recuperation) give at a rate of 1–2 mg/20 lb, 5–8 ml in an adult doe, repeated in 12–24 hours.

- To induce labor (parturition) before 144 days. Dex is the drug of choice for increasing the chance of live kids when inducing labor/parturition before 144 days. Dex does *not* bypass the stages involved in fetal lung maturation (production of surfactant) like prostaglandins (e.g., Lutalyse) do. Slower to work than prostaglandins with parturition taking 48–96 hours.

- In conjunction with prostaglandins when inducing labor (parturition) to hasten maturation of preterm or possible preterm fetal organs and tissues, particularly the lungs (to reduce respiratory distress syndrome [RDS]) , but also the cardiovascular, respiratory, nervous, and gastrointestinal systems.

- Combined with thiamine (B1) to help reverse brain swelling associated with polioencephalomalacia.

- Used in combination with flunixin meglumine to treat *E. coli* septicemia.

- Treatment of conditions where the immune system is destructively hyperactive. Higher doses are required to actually suppress the immune system.

- Blood calcium reduction (in medical conditions where blood calcium is dangerously high treatment is needed to reduce levels to normal).

Do not use in combination with medications of the NSAID class (e.g., aspirin, phenylbutazone, etc.). This combination can lead to bleeding in the stomach or intestine.

Withdrawal, Meat: 14 days. **Milk:** 72 hours.

Dexamethasone is an adrenal corticosteroid, a member of the glucocorticoid class of hormones. Unlike anabolic steroids used in sports medicine, these are "catabolic" steroids. Instead of building the body up, they are designed to break down stored resources (fats, sugars and proteins) so that they may be used as fuels in times of stress. Cortisone is a related hormone. Glucocorticoid hormones are produced naturally by the adrenal glands. Dexamethasone does not have activity in the kidney leading to the conservation of salt. This means that the classical side effects of steroid use (excessive thirst and excessive urination) are less pronounced with this steroid than with others.

Pharmacology: Like other glucocorticoids, dexamethasone acts on the metabolism (in particular on the carbohydrates) and it inhibits the adrenal cortex through feedback on the hypothalamus and pituitary; it also has strong anti-inflammatory and immunosuppressive action.

Dexamethasone use is likely to change liver enzyme blood testing and interfere with testing for thyroid diseases.

Dexamethasone is approximately 10 times as strong as prednisone/prednisolone. Its use requires certain precautions: If given for more than 24 hours, an antibiotic must be used, as Dex suppresses the body's natural immunity and leaves the animal open to infection.

Predef 2X® (Extralabel, Rx)

Isoflupredone acetate 2 mg/ml

0.3–0.7 ml/100 lb IM, *one time only.*

Like other corticosteroids (see dexamethasone above) it should not be used in pregnant does, as it can cause premature labor.

A synthetic corticosteroid. Used for treating ketosis, and in conjunction with other medications for treating allergic reactions and part of a combo treatment for shock.

Note: Dexamethasone is stronger and the better choice for deliberate induction of labor.

Oxytocin (Extralabel, Rx)

20 USP units per ml

Dosage: 10–20 units IM (0.5–1 ml) given within 24 hours after kidding is helpful in expelling retained afterbirth.

To control post-extraction cervical and uterine bleeding after internal manipulations (e.g., fetotomy, etc.), 5–10 units IV; repeat SQ in two hours.

> **Tip:** When using oxytocin to augment uterine contractions during labor, be absolutely certain that the doe is *completely dilated* and the kid is in *normal presentation*. Causing hard contractions when the fetus is not presented correctly or the cervical os is not open can irreparably harm the doe! Use with caution.

Used in minute amounts (0.1–0.3 ml) to stimulate milk let-down. Some dairy farmers and veterinarians often manage *E. coli* and *Staph* mastitis by administering small amounts of oxytocin and completing milk-out several times a day.

Metritis (uterine infection) 5–10 units IM three to four times a day for two to three days.

Withdrawal, Meat: 0 days. **Milk:** 0 days.

Prostaglandins (Extralabel, Rx)

Lutalyse® (2 mg/ml)

Dosage:

To induce heat: 0.5–1.0 ml IM to bring an ovulating adult doe into standing heat (65–75 hours after injection she will be ready to breed).

To synchronize does: 8 mg/0.75 ml Day 4 of cycle and again in 11 days.

As an abortifacient: 5–10 mg/1–2 ml.

To induce labor: 0.5–2.0 ml to bring a doe into active labor (induce parturition), with delivery in 24–57 hours. A higher dose (up to 5 ml) is reported yield a more predictable interval from injection to delivery (32 hours).

This, with careful consideration, can be used as a management tool, to terminate accidental pregnancies and to allow the owner to be on hand during freshening, as well as to aid in choosing a convenient time for both owner and buck's owner to breed the does, and, not infrequently, to get an otherwise unbreedable doe pregnant.

Not the drug of choice for induction of parturition before Day 144 of gestation for delivery of live kids. Prostaglandins bypass the steps necessary to the production of fetal lung surfactant. Before Day 144, use dexamethasone.

If doe has an unwanted breeding, wait 11 days and then give injection of Lutalyse, 2 ml IM.

Can be used after kidding to control excessive uterine bleeding.

Withdrawal, Meat: 1 day. **Milk:** 24 hours.

Warning: Pregnant women and individuals with asthma or bronchial disease should handle this product with extreme caution. If skin is accidentally exposed, wash off immediately.

Dopram, Dopram-V (Extralabel, Rx)

Respiratory and cerebral stimulant, used after a difficult birth or c-section.

Dosage and Route of Administration: Placing 10–20 drops of the solution under the tongue of the kid will produce an effect within seconds, in many cases. Inject at 0.5 mg/kg IV.

Avoid overdosing because it will cause respiratory alkalosis due to depletion of CO_2 in the blood.

Respirot (Extralabel, Rx)

Crotetamidum 75 mg, Cropropamidum 75 mg/ml

A respiratory stimulant.

Dosage: Administered buccally (inside the cheek) or nasally at 6–18 drops buccal or nasal. (One ml is about 36 drops.)

Avoid overdosing because it will cause respiratory alkalosis due to depletion of CO_2 in the blood.

Used to initiate or stimulate respirations following dystocia or cesarean section in newborns. A respiratory stimulant with a lower convulsive threshold than other analeptic agents.

Doxapram (Dopram) is a respiratory stimulant and not a reversal agent per se; however, it has been used to partially antagonize the respiratory depression produced by barbiturate anesthesia.

Xylazine (Extralabel, Rx)

Rompun®

Sedative, analgesic.

Used for disbudding in kids, or with ketamine for anesthesia.

Dosage: 0.001–0.002 ml IM for disbudding. Can be diluted with sterile water.

Use caution: It is better to under dose than to overdose.

Withdrawal, Meat: 5 days. **Milk:** 72 hours.

IV/SQ Solutions

Lactated Ringers (L.R.) (Usually Rx; occasionally OTC)

To correct fluid and electrolyte deficits and mild acidosis. *Do not use in the treatment of lactic acidosis.* Lactic acidosis occurs due to the build up of lactic acid. Adding lactated ringers will increase the lactic acid levels.

Use Normal Saline 0.9% SC (sodium chloride) to re-hydrate in the case of lactic acidosis. Use in emergency to re-hydrate animal via SQ injection or drip. *Peritoneal (IP) injection of fluids is not recommended in goats.* 1000 ml for a dehydrated adult doe (comes in 250/500/1000 ml bags). Add no more than 50 ml of 50% dextrose to the 1000 ml bag/vial.

Pharmacological Effects: Expands circulating blood volume by approximating the fluid and electrolyte composition of the blood. Lactate is metabolized by the liver and converted to bicarbonate, which aids in the correction of mild acidosis.

Active ingredient(s): Each 100 ml of sterile aqueous solution contains:
Sodium chloride	600 mg
Sodium lactate	310 mg
Potassium chloride	30 mg
Calcium chloride $2H_2O$	20 mg
Water for injection	q.s.

Milliequivalents per liter:	
Sodium	130 mEq/L
Potassium	4.0 mEq/L
Calcium	2.7 mEq/L
Total osmolar concentration	276 mOsm/L

Warm solution to body temperature and administer 1–2 ml/lb of body weight, or as determined by condition of animal, at a rate of 10–30 ml per minute under strict asepsis. May be repeated as necessary.

Precaution(s): Store at controlled room temperature between 2–30° C. (36–86° F.). Contains no preservatives. Partial bags maybe be refrigerated for 1–3 days, otherwise use entire contents when first opened.

Normosol-R (Usually Rx; occasionally OTC)

All-purpose replacement fluid used to correct fluid and electrolyte deficits and metabolic acidosis.

Some opacity of the plastic due to moisture absorption during the sterilization process may occur. This is normal and does not affect quality or safety.

CMPK (Extralabel, Rx)

For treating hypocalcemia, hypomagnesemia (grass tetany), and other conditions associated with calcium, magnesium, phosphorus and potassium deficiencies.

Dosage: 50 ml/80–100 lb, IV or SQ. (See p. 54–55 for SQ use.)

May cause severe inflammation when given SQ per mfg. Use contents or dispose of remainder.

5% Dextrose & Lactated Ringers (Usually Rx; occasionally OTC)

Dosage: Warm to room temperature and administer approximately 40 ml/10 lb SQ twice daily. For intramuscular use, dilute to 5% solution by taking 1 part 50% dextrose and 9 parts sterile saline. May also give orally.

Combo of dextrose and electrolytes gives fast energy to weak or dehydrated kids. Also to correct hypoglycemia (low blood sugar), supply water, or treat high potassium.

Saline Solution (Usually Rx; occasionally OTC).

Normal Saline 0.9% sodium chloride 4.5 gm/500 mL

To correct deficits of sodium and chloride and treat metabolic alkalosis. May be used instead of Ringer's Lactate (RL) for IV or SQ infusion.

Note: A 250 ml bottle should be kept on hand (very inexpensive) to dilute other preparations for use in infusions, etc.

Can also be used to re-hydrate a mildly dehydrated animal, particularly a kid. Inject SQ or IV drip; *peritoneal (IP) injection of fluid not usually recommended in goats.*

Calcium Gluconate 10% (Extralabel, OTC)

Calcium gluconate 100 mg/mL

5–15 mg/ml IV slowly to effect over 10-minute period. Monitor heart rate; stop treatment if bradycardia develops.

Note: Calcium chloride 10% solution is effective, but is extremely caustic if given extravascularly, but it is three times as potent, so use one-third 0.15–0.5 ml/kg. Don't confuse the two.

Calcium Gluconate 23% (Extralabel, OTC)

Calcium borogluconate 23% solution

Withdrawal, Meat: 0 days **Milk:** 0 days

Warm to room temperature and administer slowly. Store at a controlled room temperature between 59–86° F. No preservative; use completely when opened. Do not use if cloudy or contains a precipitate.

Oral Antibiotics/Antibacterials

Sulfonamides Drug Family: Folic Acid Inhibitors

Sulfonamides are broad-spectrum antimicrobials that inhibit both gram-positive and gram-negative bacteria, as well as some protozoa, such as coccidia. Sulfonamides are widely distributed throughout the body, and cross the placenta (known to cause birth defects in pregnant mice and rats that were given high doses).

Resistance of animal pathogens to sulfonamides is widespread as a result of more than 50 years of therapeutic use, which limits their effectiveness.

Bacteriostatic.

Trimethoprim/Sulfamethoxazole
Trimethoprim/Sulfadiazine (Extralabel, Rx)

Brand names: TMP/SDZ, TMP/SMX, TMP/SMZ, Tribrissen®, Bactrim®, Septra®, Septra® DS, Cotrim®, Cotrim® DS, Di-Trim® (30 mg–960 mg tablets), Sulfamethoprim®, Sulfatrim®, Sulfatrim® DS, Uroplus® DS, Uroplus® SS.

A broad spectrum antibiotic and sulfa combination with a wide spectrum of activity against gram negative and positive organisms. For scours, pneumonia and miscellaneous other infections. *There is some question about the oral use of trimethoprim in ruminants. Some believe that it may be significantly degraded in the rumen.*

Dosage: 30 mg/kg (665 mg/50 lb) twice daily. The most common tablets are 960 mg = one 960 mg tablet per 70–75 lb twice daily.

Contraindications: Untreated dehydration. Not approved for lactating animals in the US.

Withdrawal, Meat: (cattle) (Canada) 3 days for T/sulfadiazine, 10 days T/sulfadoxine. **Milk:** (cattle) (Canada) 96 hours for T/sulfadiazine or T/sulfadoxine.

Sulfaquinoxaline (20% solution) (OTC)

Coccidia treatment.

Veterinary products/brand names:
34.4 mg per mL(Liquid Sulfa-Nox)
200 mg per mL (Sulfa-Q20%; generic)
286.2 mg per mL (Sulquin 6-50)
340 mg per mL (34% Sul-Q-Nox)

Dosage: 6 mg/lb daily for 3–5 days. Drench: 2 ml/50 lb orally for 3–5 days. In water: 1 oz/10 gallons (animals will need to drink 1 quart of solution for each 25 lb of body weight to get an effective dose).

Sulfaquinoxaline is minimally absorbed systemically and is referred to as an enteric sulfonamide.

Withdrawal, Meat: 10 days.

Albon® & Di-Methox® (Extralabel, Rx or OTC)

Sulfadimethoxine

For coccidia and scours.

Dosage: 25 mg of sulfadimethoxine per pound (55 mg/kg) Day one, 12.5 mg per lb (27 mg/kg) Days 2–5 *and see below*.

Also used for pneumonia and miscellaneous bacterial infectious processes. (See Albon Injectable-40% information under Injectable Antibiotics.)

Withdrawal (boluses), **Meat:** 7 days. **Milk:** 60 hours.

Albon® (Albon 12.5% solution) Dose rate of the 12.5% oral solution is 10 ml/50 lb on Day 1, then 5 ml/50 lb Days 2–5.

Albon® (Albon boluses: 5 g and 15 g. Dose rate: 1 g/ 40 lb on Day 1, then reduce dosage by half for Days 2–5.

Albon® (Albon soluble powder) in 107 g packets. Labeled for drinking water. One 107 g packet medicates 50 gallons of water. As a drench, dissolve one packet (107 g) in two cups of water, dose at 13 ml/100 lb on Day 1 and 6.5 ml on Days 2–5, or, 1.5 ml per 10 pounds on Day 1 then 3/4 ml on Days 2–5 (keep refrigerated).

Biosol® (Approved for non lactating goats, OTC)

Neomycin

Dosage: 5 mg/lb PO once a day.

Withdrawal, Meat: 3 days

When combined with immediate removal of causative agent (e.g., too much milk, too much grain) stops scours caused by overfeeding, or non-bacterially induced gastritis (intestinal irritation). It will not cure scours caused by coccidia, *E. coli*, etc., unless combined with further antibiotic therapy (it will stop the scours temporarily, but they will continue to return until the causative organism is removed from the system.)

Miscellaneous Oral Products

Electrolytes

When picking an electrolyte (not all electrolytes are the same), read the labels. Choose an electrolyte that contains:

- Sodium—4–5 g per 100 g packet sodium corrects dehydration. (4–5% minimum)

- Alkalizing agents—can be either bicarbonate, sodium citrate, sodium acetate or a combination. (12 g, not usually listed but offers 110–160 mg/L) Corrects acidosis.

- Potassium—2–3 g per 100 g packet. Replaces electrolyte losses. (2–3% minimum)

- Chloride—4–5 g. Corrects acidosis. (4–5% minimum)

- Glycine—3–6 g. Helps water absorption. (3–6% minimum)

- Dextrose/Glucose—60–70 g. An energy source. (60–70% minimum)

- Fiber—Not a critical ingredient in electrolytes, though it does absorb water and slows down the passage of nutrients through the intestines, reduces loose stools, and helps re-establish microbial fermentation.

Products (gelling agents) used include pectins, guar gum, xantham gum, and pulps like apple and citrus. If an electrolyte contains fiber, some of the glucose was sacrificed to make room for the fiber, look for a maximum of 1–2 g of fiber in a 100 g packet.

Wait 15 to 20 minutes after feeding milk before feeding an electrolyte with bicarbonate or citrate. Together, these two alkalizing agents prevent rennin and casein from clotting in the stomach, thereby causing rapid passage of nutrients through the small intestine. Milk replacers don't contain casein, so alkalizing agents won't interfere with them.

Antihistamines

Benadryl (Diphenhydramine HCl), also sold under various brand names.

NOT FOR ACUTE ANAPHYLACTIC ALLERGIC REACTION, as this is a true emergency situation and there is no time for the oral antihistamine to be ingested and take effect.

Dosage: 5 ml (teaspoon) for very young kids to 15–20 ml for adults animals, for treatment of mild allergic reaction to bite/sting, medication. Also as nasal decongestant and cough medication.

Will counteract the slower form of allergic and histamine reaction that takes place sometime in the 24-hour period after severe stress or injection of something to which the goat is allergic.

Dimetapp. A good nasal decongestant, etc., that works well for kids. Tavist-D and Chlor-Trimeton (Chlorpheniramine Maleate) have also been reported used with good results.

CMPK

Available as a gel or a drench, used as a supplemental nutritive source of calcium, phosphorus, magnesium, potassium and dextrose and for treating hypocalcemia.

Kaolin Pectin

Made from a clay and an emollient, kaolin pectin is inexpensive and is used to treat diarrhea in animals. (Avoid the newer formulation of this, which contains salicylates!)

Milk of Magnesia

The product of choice for a goat that is a little bloated or has minor diarrhea.

Propylene Glycol

A MUST to have on hand for maintaining pregnant does.

An excellent energy source as well as being glycogenic. Propylene glycol may be used in prevention and treatment of ketosis. It has an approved health claim for this use.

Dosage: 3–4 oz (90–120 ml) 2 times daily, for 2 days, and then 1–2 oz (30–60 ml) 2 times daily until animal is eating satisfactorily again.

Administer ANY TIME A PREGNANT OR EARLY LACTATION DOE GOES OFF FEED.

Can also be used as a diluent for oral administration of non-water soluble drugs, e.g., ivermectin.

Probios®, FastTrack®

Rumen inoculant. Probiotic. Contains specific, beneficial bacilli (bacteria) to keep the rumen working. Use anytime an animal is "off." Use probios and vitamin B-Complex as a first line of defense.

Yogurt containing active cultures (check the label) can be used in an emergency, but is not as specific as Probios and other products formulated for goats.

Use during and following antibiotic treatment (antibiotics kill off the good bacteria along with the bad).

Dosage: 5 g for kids; 10 g adults.

Oral Pain and Anti-inflammatory Medications

Aspirin (Extralabel, OTC)

Dosage: 50–100 mg/kg PO every 12 hours.

Withdrawal: Not approved, unofficial. **Meat:** 1 day. **Milk:** 24 hours.

Banamine® - Citatron® (Extralabel, Rx)

Flunixin meglumine paste and crumbles.

Dosage: 1 cc/100 lb. Some recommend every 36 hours, but one study showed that it leaves the system after approximately 22 hours.

> Oral administration of Banamine is as effective, with comparable absorption, to administration either IM or IV.
> *Acta Vet Scand* 44(4): 153–59

Long-term use not recommended (possible stomach ulcers, kidney problems).

Contraindications: Gastric ulcers, concurrent kidney disease, untreated dehydration, other NSAID use.

Withdrawal, Meat: 10 days. **Milk:** 72 hours.

Bute/Butazolidine (Extralabel, Rx)

Phenylbutazone

NSAID for pain, especially of the joints and muscles. Can cause stomach upset and ulcers, with long-term, daily use. *Do not use in lactating animals.*

Dosage: 4.5–10 mg/lb PO, once daily.

Not approved for food-producing animals. In 2003 the FDA prohibited extralabel use of phenylbutazone in female dairy cattle 20 months of age or older on evidence that it will likely cause an adverse event in humans.

Withdrawal, Meat: 10 days. **Milk:** 72 hours.

Ketoprofen

Dosage: 1.5–2.0 times the human dose, once daily.

Withdrawal: Not approved, unofficial. **Meat:** 7 days. **Milk:** 24 hours.

Ibuprofen

Dosage: Give adults twice the human dose two times daily, with food to protect the stomach.

Other Miscellaneous Products

Activated Charcoal, ToxiBan

An emergency product for use when a goat has swallowed a toxic substance. It is believed to deactivate up to 60% of poisons. ToxiBan is a charcoal-kaolin suspension.

Betadine® Solution

Povidine-iodine. Excellent for irrigating wounds and cleaning abscesses. Use 1 part to 10 parts sterile water. Effective against a broad spectrum of organisms.

Betadine® Surgical Scrub

Povidine-iodine based sudsing cleanser. Use to coat a washed hand or sterile gloved hand when doing a vaginal check on a kidding doe. For sanitizing skin surfaces for surgery, as well as for cleansing wounds or abrasions.

Blood Stop Powder

Styptic powder used to stop bleeding from hoof-trimming, loss of scurs or other injuries.

Hydrogen Peroxide 3%

Mild disinfectant. Useful for irrigating and cleansing abscesses and open wounds before applying antibiotic preparations.

An oxidizing agent, hydrogen peroxide combines with organic material in wounds to break down large molecules and liquefy solid matter. When hydrogen peroxide solution contacts organic material containing catylase, oxygen is liberated rapidly enough to cause effervescence, which provides a mechanical cleansing effect. This action is especially useful in grossly contaminated wounds or open cavities that have difficult-to-penetrate areas, such as jagged wounds with extensive tissue damage. The antiseptic or germicidal action is brief and ends with the completion of oxygen release, evidenced by the subsidence of visible foaming.

Note: Never inject hydrogen peroxide.

Iodine 7%

Apply liberally to naval cords of newborns. Keep in small container and discard after dipping 3–4 kids (the active ingredient is used up). Betadine, peroxide and milder disinfectants do not provide adequate disinfection for naval cords; use only 7% iodine.

Nolvasan

Chlorhexidine disinfectant. Effective against at least 60 different bacteria, fungi, yeasts and viruses. Because it binds to the skin, its effect can last for as long as two days.

Triple Antibiotic Ointment

Neomycin and Polymyxin B Sulfates and Bacitracin Zinc Ointment USP.

Salve to treat minor cuts and abrasions. Prevents infection and aids in healing.

Neosporin® (Gramicidin; Neomycin; Polymyxin B)
Polysporin (Bacitracin Zinc; Polymyxin B Sulfate)

Antibiotic ointment for quick healing of scratches, scrapes and irritated skin.

Red Cell

Yucca-flavored vitamin/iron/mineral supplement

Note: Give goats the horse formulation.

Give 1 cc/20 lb every 12 hours for anemia.

Terramycin Eye Ointment

Antibiotic (oxytetracycline) eye ointment. For treatment of pinkeye (contagious keratoconjunctivitis), apply three times daily. *Mycoplasma* and *chlamydia* are believed to be the most common causes of pinkeye in goats. Both respond to oxytetracycline; neither respond to penicillins.

Zinc Oxide

Good for treating and preventing urine scald in breeding bucks and in wethers that have undergone surgery for urinary stones that leaves them incontinent.

Zinc Sulfate

Dr. Naylor Hoof 'n Heel

Zinc sulfate and sodium laurel sulfate

For treating foot rot, clean hoof and treat once or twice a day until healed.

Adapted from *Ruminations* #27–29

Mastitis Detection and Treatment

California Mastitis Test (CMT)

A must for those with milking does. It should be used routinely at least once a month to detect sub-clinical mastitis before it damages the udder. The most common dairy goat mastitis, *staph aureus*, may cause permanent damage. Anytime a change is seen, such as a lessening of production in one side or the other of the udder or chunks in the milk, a CMT should be run immediately.

Take milk, mid-milking, from each side of the udder separately (clean utensil before taking milk from each side) and place one teaspoon of milk in well of paddle. Add an equal amount of CMT liquid and swirl them together.

The degree of coagulation (thick or thin) indicates the level of severity of the infection. If infection is present, begin treatment of the infected side at once, with the advice of an experienced dairy goat breeder or knowledgeable veterinarian.

Udder Infusions

Administer infusions—4 or 5 infusions 12 hours apart, or 2 infusions 24 hours apart (for less serious infections)—accompanied by systemic penicillin or oxytetracycline to correct the sub-acute form of udder infection. Milk from the affected half of the udder should be cultured to determine the correct medication to use. Before treating, collect a sample into a sterile container and freeze to isolate the organisms. The numbers of samples positive for *E. coli* may decrease after freezing. Samples that have been frozen cannot be used for somatic cell counts.

Dry Cow Udder Infusions

Tomorrow®/Cefa-Dri® (long acting cepharin)

Dry Clox® (Rx) and **Orbenin®** (Rx) (both cloxacillin)

QuarterMaster xz (dihydrostreptomycin & penicillin G).

FDA-Approved Drugs for Intramammary Use in Non-lactating Cows

ACTIVE INGREDIENT	TYPE	MILK W/H	MEAT W/H	NAME	MFGR/MKTR
Amoxicillin	Rx	60 hrs	12 days	Amoxi-Mast	Schering-Plough
Cephapirin (benzathine)	OTC	72 hrs Postcalving	42 days	Cefa-Dry/Tomorrow Intramammary Infusion	Fort Dodge/Wyeth
	OTC	96 hrs	4 days	Cefa-Lak/Today Mammary Infusion	Fort Dodge/Wyeth
Cloxacillin	Rx	Not allowed	30 days	Dry-Clox Intramammary Infusion	Fort Dodge/Wyeth
	Rx	48 hrs	10 days	Dariclox	Schering-Plough
	Rx	Not allowed	28 days	Orbenin DC	Schering-Plough
	Rx	72 hr Postcalving	30 days	Boviclox	Northbrook Laboratories, Ltd.
Dihydrostreptomycin sulfate	Rx	96 hr Postcalving	60 days	Quartermaster Dry Cow Treatment	West Agro, Inc.
Hetacillin (potassium)	Rx	72 hr	10 days	Hetacin-K Intramammary Infusion	Fort Dodge/Wyeth
Novobiocin	OTC	72 hr Postcalving	30 days	Albadry Plus Suspension	Upjohn Co.
	OTC	Not allowed	30 days	Drygard	Upjohn Co.
Penicillin G (procaine)	OTC	72 hr Postcalving	30 days	Albadry Plus Suspension	Upjohn Co.
	OTC	72 hr Postcalving	14 days	Go Dry; Masti-Clear	G.C. Hanford Mfg. Co.
	Rx	96 hr Postcalving	60 days	Quartermaster Dry Cow Treatment	West Agro, Inc.

Information from Food Animal Residue Avoidance Databank (FARAD), current as of September 2008

Anthelmintics (Dewormers)

DRUG	APPROVAL	DOSAGE	ROUTE	MEAT W/H	MILK W/H
Ivermectin	Extralabel				
Ivomec® drench		0.13 mg/lb (0.3 mg/kg)	PO	140 days	9 days
Ivomec® 1%		0.13 mg/lb (0.3 mg/kg)	SQ	56 days	50 days
Doramectin	Extralabel				
Dectomax®		0.13 mg/lb (0.3 mg/kg)	SQ	56 days	40 days
Eprinomectin	Extralabel				
Eprinex®		0.22 mg/lb (5 mg/kg)	PO	NA	NA
Moxidectin	Extralabel				
Quest®, Cydectin®		0.09 mg/lb (0.2 mg/kg)	SQ	30 days	Do not use

Note: Moxidectin has been found to be more effective when given SQ.

Albendazole	Extralabel				
Valbazen®		4.5 mg/lb (10 mg/kg)	PO	7 days	120 hours

Note: Do not use in does during the first 30 days after breeding; may cause birth defects.

Fenbendazole	Approved				
Safeguard®					
Panacur®		4.5 mg/lb (10 mg/kg)	PO	4 days	120 hours
Oxfendazole	Extralabel				
Synanthic®		4.5 mg/lb (10 mg/kg)	PO	10 days	120 hours
Levamisole	Extralabel				
Levasole®		3.6 mg/lb (8 mg/kg)	PO	4 days	4 days
Morantel Tartrate	Approved				
Rumatel®		0.5 mg/lb (10 mg/kg)	PO	3 days	3 days

Anticoccidial Products

Amprolium (Corid®) (Extralabel, OTC)

Dosage: On packet or bottle—can be administered in the drinking water or as a drench.

To treat clinical cases, use 11.3–23 mg/lb (10 mg/kg) daily in feed or water for 5 days.

To prevent coccidiosis, use 2.3 mg per pound (5 mg/kg) body weight daily for 21 days.

An alternative to sulfa for treatment of coccidia, though the sulfas are a better choice for treatment of coccidiosis or severe coccidia problems.

Amprolium is an antagonist of thiamine (vitamin B1), one of the essential vitamins, and thus interferes with the metabolism of the parasites.

Decoquinate (Deccox) (Approved for goats, OTC)

Do NOT feed to goats producing milk for food.

Dosage: For prevention, at 0.5 mg/kg. Feed for at least 28 days during periods of exposure to coccidiosis or when it is likely to be a hazard.

Lasalocid (Bovatec) (Extralabel, OTC)

Dosage: For control, at 1 mg/kg.

Monensin (Rumensin) (Approved for goats, OTC)

Dosage: For prevention, at 1.2 mg/kg 0.4. For control, at 0.8 mg/kg.

These products have varying levels of anticoccidial activity at their label doses. A good coccidiostat can break the life cycle of parasites so they can't reproduce and spread infection, stopping them before they multiply and create massive gut damage.

Coban 60 (Approved for goats, OTC)

Monensin Type A containing 60 g/lb. Must be mixed in feed at a rate of 20 g/ton of feed. No meat withdrawal. Prohibited in lactating goats.

When mixing and handling Monensin Type A medicated articles, use protective clothing, impervious gloves, and a dust mask. Operators should wash thoroughly with soap and water after handling. If accidental eye contact occurs, immediately rinse thoroughly with water. Do not allow horses, other equines, mature turkeys, or guinea fowl access to feed containing monensin. Ingestion of monensin by horses and guinea fowl has been fatal.

Natural Care and Home Remedies

Goat keepers have used herbal remedies and home remedies for probably as long as goats have been domesticated. Pharmaceuticals are a modern invention and in some cases came out of herbal remedies. Examples of this include foxglove, from which digitalis came, and white willow bark, which was used by Hippocrates in the 5th century BCE, and contains salicin, the compound from which aspirin is made.

Natural care, including herbal remedies, is often the best choice for those who use or sell their goat milk. They have a legal and moral obligation to make sure that the milk contains no drug residues. Others may choose herbal care because it costs less, is considered more "natural," or because they do not have access to pharmaceuticals.

Home remedies may be considered natural care or not, depending on what ingredients are being used to treat the goat. For instance, honey for healing wounds is a form of natural care, where oxytetracyline in the eye is not.

These recipes were compiled from various sources over the years and can be used as an adjunct to modern veterinary treatments or, where appropriate, by themselves. The intent is to give goat keepers alternatives to health care management for their goats.

Home Remedies

Homemade Electrolyte Solution

Use this solution to treat a dehydrated kid or adult goat or a doe with ketosis. Have the goat drink this mixture or use a feeding tube.

 1 teaspoon salt
 ¾ teaspoon Lite Salt
 1 teaspoon baking soda
 4 oz corn syrup
 warm water to make 4 pints

Mix all ingredients well. Give a large dairy goat one pint of the solution 3–4 times daily. Give ½–1 cup, four times a day (every six hours).

Goat Magic

Most goat owners learn early to keep on hand some Nutridrench or a similar product to give their goats a boost when they are feeling down. Here is a homemade alternative that costs only pennies and can be made with ingredients usually on hand.

 8 oz hot water
 2 tablespoon molasses
 2 tablespoon light corn syrup
 ½ teaspoon salt
 ½ teaspoon baking soda

Stir well to mix the ingredients and serve hot. Goats love a hot drink, in fact many breeders provide hot water for their herds whenever the temperature drops. This helps hydrate, stimulates appetite, helps the natural immune system and eases stress.

From *Ruminations* #58

Colostrum Substitute

3 cups milk
1 beaten egg
1 teaspoon cod liver oil
1 tablespoon sugar

Mix ingredients well and, over the first few days of life, bottle- or tube feed a kid that has no access to safe colostrum, or to supplement colostrum. This provides nutrition but not the antibodies of real colostrum. It should be used a last resort when safe goat, sheep or cow colostrum is unavailable. Make sure to keep the kid warm, clean and dry.

Bloat Prevention

Provide baking soda, free choice.

Bloat

1 cup fizzy soft drink, *not* diet; it must be the real thing! Even soda water will do; ginger beer is best. Walk the animal round until it burps and farts.

Irene Ramsay
New Zealand

Blood Stop

A handful of cobweb applied to the area will stop bleeding.

Colic

In an emergency when immediate help is unavailable, give 300 ml milk of magnesia mixed with six soluble aspirins. This is an adult dairy goat dose.

Irene Ramsay
New Zealand

Honey for Wound Care

Honey is a treatment that was frequently used prior to the discovery of penicillin. Recent studies have shown it to be effective in burns, wounds and even methicillin resistant *staph aureus* (MRSA). Completely cover the wound with honey. The natural hydrogen peroxide helps with healing and the texture of the honey keeps out the bad bugs.

Even more beneficial is Active Manuka Honey, available from and licensed as a medicine in Australia and New Zealand. The bees make the honey from flowers of the tea tree, which is well-known for its healing properties.

The disadvantages of honey as a treatment for a goat are attraction of flies and licking by other goats, which can limit its efficacy.

Hypomagnesemia

Milk of magnesia, double human dose for weight. The initial symptoms are similar to milk fever. The difference is that in magnesium deficiency the eyeballs twitch like crazy. It won't harm a goat with milk fever to give the milk of magnesia if you aren't sure, but you won't cure a hypomagnesemia with milk fever treatments, so doing both can cover you.

Irene Ramsay
New Zealand

Indigestion and/or Not Chewing Cud

Mix 1 teaspoon baking soda and ½ teaspoon ground ginger 20 ml water and drench.

Irene Ramsay
New Zealand

Labor Energy Boost

For a goat that is tiring during a long birth, give a strong cup of coffee, with or without sweetener.

Mastitis Treatment

- Hot packs to the udder
- Give aspirin for inflammation

Painkiller

To 1 cup warm water add:

1 tablespoon vinegar
1 teaspoon honey or sugar
1 good pinch each of salt, baking soda, and ground ginger.

Give as required. If the animal has not been able to sleep due to pain, it may fall asleep for some hours about 20 minutes after swallowing the first dose. Can also be used for oral rehydration. (It was originally developed, without the ginger, for oral rehydration of cholera victims when there wasn't enough IV equipment to go round.)

Pinkeye Treatments

- Try a drop of oxytetracycline directly in the eye. It is very effective on susceptible bacteria.
- Use a drop of port wine in the eye to treat pinkeye.

Stealing a Cud

One method to get a sick goat to ruminate again, after trying probiotics or yogurt, is to steal a cud. Sometimes this is nearly impossible to do, and you can get bit.

An alternative method is to use a drench gun filled with warm water to flush out the mouth of a healthy goat just as she brings up a new cud. Your partner can catch the run-off in a bowl. Then suck up the liquid with a turkey baster or a syringe and drench the sick goat with it.

From "Health and Science News," *Ruminations* #58

Irene Ramsay's Rhododendron (and Other Plants) Poisoning Remedy

15 ml rennet
15 ml milk of magnesia (MoM)
5 ml brandy

Mix together. This is the adult dose!

For kids under four months, give 5 ml each of rennet and MoM and 2 ml brandy. For kids over four months give 10 ml of each and 5 ml brandy. The action of the rennet is to neutralise the toxins from the rhododendron. Milk of magnesia is useful for a number of stomach upsets in goats (as well as in humans). Its action is twofold: It lines the gut, and it regulates the pulsing of the gut (peristalsis), which often gets out of kilter with poisoning or colic.

The brandy works, but I haven't yet found out why. It is a fortified spirit so has a high alcohol percentage and you don't need much. Alternatively you can use sherry, which is a fortified wine. Both are made from grapes, and work better for medicinal purposes than spirits or wines made from grain or other substances.

It is usual for goats with rhododendron poisoning to vomit rather spectacularly. For this reason it is difficult to drench them with an antidote because it is easy to cause aspiration or inhalation pneumonia, which aren't treatable in the farm situation. I like this recipe because the amounts are small, and you can take 15 minutes over the drenching, a ml or two at a time between sickies, if you have to.

One dose is usually sufficient. It also is important to keep the goat warm, but not in the sun, and out of the wind. Keep a bucket of *clean* fresh warm water available for the animal to drink. Once the goat is feeling better, offer a handful of various weeds like yarrow, cleavers, dock, prairie grass or twitch (couch grass), some green pine or fir needles, tree lucerne (tagasaste), willow, and some good plain hay or straw. Don't give much at a time in case the goat vomits again.

Lorraine's Rhododendron Poisoning Recipe

(Quantities do not need to be exact)

¼ cup cooking oil
½ cup strong cold black tea (6 to 8 tea bags, removed)
1 teaspoon ground ginger
1 teaspoon baking soda

Mix together and drench the goat.

Oil lines the stomach, preventing more poison from going into the system, tea is the antidote, ginger relieves pain, and baking soda helps bring up the gas.

Scours
- Roughage, such as blackberry leaves, madrone or salal
- Pepto Bismol (Contains salicylates; do *not* give to pregnant does)

Ringworm
Apply Absorbine Jr. Original Pain Relieving Liquid to the ringworm twice a day until it disappears.

Udder Lumps
10 ml Preparation H cream (not ointment)
1.5 ml liquid DMSO

Wearing gloves, mix ingredients well and put in airtight container. Twice a day, massage a small amount into the udder, always wearing gloves. Use for three to four days in conjunction with an antibiotic infusion. Discard milk for 72 hours or more.

Amy Kowalik
Pipe Creek, Texas

Getting Goats to Take Medicine
Mix medications, such as ammonium chloride into Jello Jigglers, and the goats will like taking them!

Do Nothing
When I was growing up my mother's favorite remedy was "Ignore it and it will go away." This often works with minor maladies. For instance, if a goat has a mild case of diarrhea, perhaps he or she simply has a temporary digestive disturbance. If no other symptoms are present and it does not persist, you probably can ignore it, and it will clear itself up. The same is true with a minor fever—the body's own disease-fighting mechanism.

Herbal Remedies

Birthing Tonic

Mix equal parts red raspberry leaf, powdered cloves, ground ivy and powdered thyme with molasses and form into balls. Roll in slippery elm powder. Give to doe 3–4 weeks prior to kidding.

Tip: Always use unsulfured molasses in goat water or herbal preparations to avoid polioencephalomalacia.

CAEV Palliative

1 teaspoon–1 tablespoon turmeric powder added to hydrated beet pulp or pelleted feed.

Coccidia Prevention

Feed Douglas fir *(Pseudotsuga menziesii)* branches regularly, especially in the spring after kidding.

Immune System Stimulation

- Echinacea
- Garlic

Increasing Milk Supply

Make sure to massage and "bump" (like the kids do, but gently) the udder at the end of milking.

To assist or improve a doe's milk supply give any of the following: oats, turmeric (believed to help prevent mastitis, as well), alfalfa, barley, fennel (also a digestive aid), ginger or fenugreek.

Intestinal problems

8 parts slippery elm
1 part licorice
1 part fenugreek

Mix with molasses, form into balls and feed to goats.

- ginger (aids digestion)
- fenugreek (soothes the intestine)
- fresh plaintain leaves, calendula flower heads, nettles and comfrey leaves
- slippery elm bark powder

Kidding

- Give red raspberry leaves or nettles to a doe a few weeks prior to kidding to help with contractions.

- Provide raspberry leaves as a uterine tonic after kidding; use blackberry leaves if raspberry leaves are not available.

- 4–5 leaves of fresh ivy per hour will increase uterine contractions.

Mastitis Treatments

- Massage the udder with peppermint or eucalyptus oil in a carrier oil such as olive or almond.

- Give echinacea at the first signs of mastitis. Note that it becomes less effective if used frequently or for a long time.

- Make a poultice of comfrey leaves and calendula flowers by blending and mixing with flour. Put on cotton cloth and apply hot to the udder. Treat frequently.

- Make a rosemary *(R. officinalis)* infusion by adding 2–4 teaspoons of fresh or dried rosemary to a cup of boiling water. Steep for 10 minutes, then strain and use in a poultice.

Milk Gruel for Sick Goats

20 oz barley flour
20 oz oat flour
2 oz slippery elm bark
2 oz white willow bark
2 oz marshmallow root
2 oz arrowroot powder
1 oz dill seed

Mix with liquid colloidal minerals and/or molasses and form into small balls or feed as a gruel.

Pain Relief

- 1 tablespoon chopped valerian root steeped, covered, in 1 cup hot water for 20 minutes and fed to goat.

- 1 tablespoon chopped catnip steeped, covered in 1 cup hot water for 10 minutes and fed to goat.

- Willow branches (don't overdo it, as they can cause stomach upset).

Parasite Control

- black walnut hulls
- clove buds
- false unicorn root
- garlic bulb
- hyssop leaf
- myrrh gum
- peppermint leaf
- pumpkin seeds
- wormwood
- chlorophyll
- elecampane
- fenugreek seeds
- gentian root
- male fern root
- oregano
- prickly ash bark
- turmeric root
- yellow dock

Some people dust their animals with diatomaceous earth (DE). DE is the fossilized remains of the diatom, a single-celled aquatic plant that lived about 20 million years ago. It dehydrates soft-bodied insects. When added to feed, food grade DE is also claimed to be a good internal anti-parasitic.

Pinkeye

Make a tea with eyebright (*Euphrasia officinalis*), using 1 heaping tablespoon per cup of water and steeping for 20 minutes. Make a compress with a clean soft rag and apply to the eye several times a day.

Reducing Milk Production

Put a paste of 1 tsp dried sage (*Salvia spp.*) mixed with water on the udder. Feed dried sage with molasses in grain.

Retained Placenta

Give a handful of English ivy (*Hedera helix*) leaves at kidding to contract the uterus and prevent retained placenta.

Skin and Coat Health

Feed kelp, free choice.

Urine Scald

- Propolis, beeswax and shea butter can be made into a barrier salve.
- Aloe and vitamin E can help with healing.

Urinary Stones

⅓ cup apple cider vinegar diluted in 1 cup water, twice a day.

Fighting Flies in the Goat Barn

Spring and summer bring not only new kids and good weather, but flies, mosquitoes and other pests. *So what is the best way to control them?*

Insecticides should be used sparingly or not at all, as they can have negative effects on both humans and farm animals.

The cheapest thing you can do is to keep the barn dry, eliminating wet bedding at least weekly.

Fly parasites can be used as part of an integrated pest management program. These little critters are actually tiny wasps that are harmless to all but fly larvae. They lay their eggs in the larvae and the next generation of parasites eats the future flies before they hatch. The parasites are about the size of a gnat, and practically unnoticeable. They don't sting or bite humans or other animals.

Another method is the tried and true fly strips. These are quite inexpensive and can be hung throughout the barn with no adverse outcome, because they use a sticky substance, rather than an insecticide, to catch the flies.

Similar to the fly strips are fly rolls. The sticky tape is on one roller and is threaded onto an empty one a few feet away. When the exposed tape is filled with flies, it is wound onto the empty roller to expose more tape. These cost a bit more, but also require less changing.

Another method is the plastic bell-shaped traps that are filled with water and stinky bait. The flies are attracted to the smell, crawl in and drown or can't get back out. These smell bad, but they work.

I also use citronella spray, made for horses, directly on the goats when the flies become too aggressive. This is a natural fly repellent and gives the goats some peace from the ongoing biting.

Check the Internet and vet supply catalogs for these products before the flies get too bad. Remember that the fly life cycle is about two weeks long. The larvae develop during the first week and pupate the second week before emerging as adult flies.

From "Health and Science News," *Ruminations* #39

Alternatives: Herbal Remedies

by Donna Geiser

The whole world is Mother Earth's natural pharmacy. No matter where we live there are plants that provide the needed remedies for ailments in animals and humans. This, of course, is contingent on the containment of urban sprawl. Because we have built homes and cities where once native plants grew, they have become increasingly difficult to find. The following are some plants that are still accessible and a couple of remedies that I have found useful in my herd.

1. **Artemisia.** Common plants are southern wormwood, mugwort, and sagebrush. The flowers and leaves are used as a natural wormer. It sometimes can be found in combination with other herbs in capsule form. Some people use artemisia for worming and as a coccidia preventive. Since we have an abundance of sagebrush here on the ranch, the goats have access anytime they feel they need it.

2. **Dandelion.** Yes, the common dandelion that grows abundantly in lawns is actually a very good herb for cleansing the liver. The body—human and animal—accumulates toxins, especially during the winter months. We've all heard of using the young tender leaves in salads, that's the gourmet way! You'll be doing two things at the same time, giving the goats the needed greens and getting exercise pulling dandelions from the lawn!

3. **Mint.** This plant is easily grown anywhere that moist, well-drained soil is available. It is quite an invasive plant so I advise planting where it can spread freely. Mint is wonderful for the digestive system and goats will nibble on it as needed.

4. **Bryonia Alba (30C).** This is a great homeopathic remedy for goats that have a cough that seems to be aggravated when in motion. I recently had a yearling doe who coughed every time she moved with the herd or tried to run and play. She had no runny nose, temperature, or outward symptoms that would indicate a serious ailment. I started by giving her a dose in the morning and evening for three days, then just once a day for another three or four days. It completely cleared up and has not been a problem since. Sometimes the change in weather from season to season will exacerbate congestion in the respiratory system.

As we begin to ready ourselves for daily milking chores, show season, and promotion of our favorite breeds in marketing strategies, it would be great if we could each learn about one new alternative remedy that can be passed along to others.

From *Ruminations* #53

Alternatives: Teeth

by Donna Geiser

Teeth. Not my teeth. Not your teeth. Goat teeth. Although I usually write about herbal and homeopathic remedies, I wanted to bring to light some issues regarding teeth. This is an important yet overlooked area of our goats' health. I think we all take for granted that their teeth will function correctly during their lifespan. In most instances this is true, however when a problem arises we should be prepared to deal with it.

What food contributes to good teeth? Goats need long-stemmed, pliable grasses, shrubs and hay in order for their back molars to grind efficiently. A balanced diet that includes such plant material will ensure that their teeth wear down evenly. Consumption of molasses-coated grain/feed in large quantities or a complete diet of pelletized food, without the addition of quality hay, may contribute to the creation of sharp points and uneven wearing of the molars.

Sometimes an injury to the jaw as a kid, or a congenital birth defect of the mouth will become problematic as the goat ages. When "sharp points" appear on those molars, a procedure that horse owners use is to "float" the teeth. A dental tool that looks much like a long-handled wood rasp or file is used to smooth those points down so the goat can grind its food properly. A vet who is knowledgeable about goats or a supply store/catalog may carry the tool for purchase.

How does a breeder know when a goat has a problem with its teeth? Observation is the key. I am writing about an experience with a goat I acquired a couple of years ago. She was about three years old when I noticed she wasn't maintaining her weight and her milk production was dropping rapidly. Of course I did all the normal things we as breeders do: upped her grain ration on the milk stand, wormed her, added more protein and vitamins to her diet, etc. As I watched her eat, I noticed hay sticking out of the sides of her mouth. The hay just stayed there until I pulled it out. She would expel clumps of wet smelly undigested food matter or I would find it floating in the water bucket. I knew something had to be done or I would lose this doe. As a breeder without access to a goat vet with the correct tools, I had to improvise. Haven't we all had an occasion where we had to substitute items or use makeshift ones?

Four of us worked together: a friend who had come over, my husband, and a neighbor taking a walk past the ranch who was commandeered to help. We found a soft block of wood to put into her mouth to keep it open. While the neighbor and husband held the doe's body, my friend held the wood block in the mouth and I went back and forth on those back molars with a wood file. We stopped and let her rest a while, then continued several more times until her back teeth did not feel sharp or rough. She took about three weeks to start eating normally. No more stuck hay in her mouth or sticking out the sides. This was not a pleasant experience. We all felt sympathy for the doe while I was filing her teeth and I certainly would not recommend doing this unless it was a last resort. Necessity was the mother of invention during this crisis.

Just recently I had a three-year-old goat I had acquired who was not chewing normally. (I have no idea why I had to be the one experiencing this phenomenon!) A "gut" feeling told me something was wrong with her mouth. Not wanting to jump to conclusions, I tried worming, probiotics and changing hay. Nothing worked. Instead of hay sticking out of her mouth, she would spit out what looked like little bales of hay. They were everywhere. They almost looked like unchewed cud. I stood behind her and gently held her jaw in one hand while I slid my other hand inside her mouth along the cheek and gum. I felt thickening of the cheek tissue and irregular teeth protruding every which way. I couldn't visualize what exactly I was feeling!

Finally, I followed up on leads for goat vets and this time was lucky enough to find one about two-and-a-half hours away. We scheduled an appointment and prepared ourselves for the worst. Although the vet had experience with goats, she had never seen anything like this. The lower molars didn't line up with the top ones. The lower molars hit the gum line of the top ones and loosened them enough that they flared out, rubbing against the inside cheek. There was a build-up of layered scar tissue on each side and teeth so jagged they couldn't be smoothed out. We found impacted food material everywhere and at that point ran out of options. We went ahead and put her down—three days before Christmas. This was some kind of malocclusion that may have been a congenital condition; none of her siblings or offspring showed any problems. Unfortunately none of the pictures I took at the vet's office turned out.

I learned from these experiences that we should always observe the eating habits of our goats, watch how they chew their food and provide them the correct types of forage so they use their grinding teeth properly. When a problem does arise with a particular animal, knowing their normal routine will help to evaluate what steps can be taken to correct the situation. The solution may be as simple as using a float to smooth the sharp points. Although I am by no means an expert in the field of goat dentistry, I hope that sharing these experiences with other breeders may help them to determine whether problems exist, before they become life-threatening. Teeth are important; not only ours, but our goats', too!

From *Ruminations* #56

Alternatives: Breeding

by Donna Geiser

Summer is ending and fall will be here soon. Cooler evenings and mornings bring the bucks' aroma floating in the air to attract the does. Ah—breeding season and the promise of spring kids. Are you ready—are your does ready?

If you've never had the occasional doe who continued to cycle and had difficulties settling, you are very fortunate. Many of us have had to deal with this problem. Sometimes cysts form on the ovaries making it difficult for the doe to conceive and/or hampering conception. This usually happens to older does that have previously freshened; however, it can happen to maiden does, too. Some breeders have tried Cyctorelin and even Lutalyse with mixed results. An alternative approach to this condition has worked well for me—I use homeopathic remedies.

The remedies are as follows: Give one dose of Apis mellifica 6C twice a day for 10 days. On the third day add a dose, once a day, of pulsatilla 30C for seven days. You will finish both remedies at the end of the 10-day sequence. In homeopathic practice, a dose can be one, two, or three of the little pellets. I usually give two and, to make giving them easier, I set them on top of a little grain in my hand.

Homeopathic remedies can be found in any good health food store and they come in a tube the size of a chapstick. You can not substitute the "C" potency for an "X", as "X" is a lower strength. Make sure you purchase the "C" strength.

You may notice vaginal discharge during or immediately after the conclusion of the 10-day program. Do not be alarmed. It is the body's natural way of eliminating matter inhibiting normal function of the uterus.

I let the doe cycle once after the completion of the remedies, then put her together with the buck on the next cycle. I also use this method as a preventive for many of my older does before breeding. Apis mellifica reduces swelling and can be used for bee/wasp stings, etc. It is a handy remedy to have on hand around the ranch or farm.

Pulsatilla can be used successfully to support the female reproductive system. I also give my does a couple of doses after kidding to support the uterus.

Using homeopathic or herbal remedies is a natural way to allow the body to heal itself.

From *Ruminations* #54

Alternatives: Pre-Kidding

by Donna Geiser

Every time I turn on the television, I see commercials advertising some new medication created to improve our lives. However, the possible side effects listed take up half the commercial. I especially like the sleeping aid with the butterfly floating through the open window. One of the side effects is "may cause drowsiness." Can you believe that? I think people who have problems falling asleep should visit goat breeders. Yes, feeding, milking, hoof trimming and pen cleaning should do the trick. Then they can crawl into bed underneath sunshine-fresh, clean, just-dried on-the-line sheets. I bet they wouldn't need to take a pill to fall asleep! Sleep would come naturally, and isn't "naturally" a part of raising goats or any animal that should be considered and investigated?

I certainly am no certified herbalist, homeopathic practitioner or licensed physician. I am just an individual who has had success in using herbal and homeopathic remedies for over 20 years—on my human and animal family. I have found the goats rebound much faster using natural alternatives . They have no pre-conceived ideas about them and no one told them to be skeptical. Goats and animals in general are very accepting of the methods chosen to treat their ailments. We all use alternatives without realizing it: baking soda, Pepto Bismol, Karo syrup, treats such as apples/carrots/raisins, and yogurt. These products have been around for a long time and are easily accessible in our cabinet or refrigerator—and they have proven successful!

Since kidding season can be a stressful time for all of us no matter how long we have been involved, I'd like to present a few remedies that I have used for many years.

Liquid Hawthorn Berries

This herb affects the circulatory system. It contains naturally-occurring chemicals that enhance the oxygen utilization in the heart muscle. Because it is in liquid form, it is easily assimilated by the body. If I have a newborn kid that just needs a jump start, I pour some on my finger and insert into its mouth. If needed, I will give another application in 10- to 15-minute intervals until I see improvement. I have been very impressed with how rapidly a kid will respond to this product. I find watching how quickly they are up and searching for mom's udder to be very rewarding.

Red Raspberry Leaves

The leaves are extensively used for their vitamin C content to support the whole reproductive system and to relax the uterus. I give some of the leaves once a day to a doe three or four days before her due date. I don't always have access to the bulk leaves, so I purchase the herbal capsules from a health food store. I either give one whole capsule to each doe, or break it open and add it to a little grain. This herb benefits first fresheners and older does due to kid. One nice thing about using herbs is that the body will eliminate any amount that is not needed.

Arnica Montana

This is a homeopathic remedy and the one I use the most. Arnica is the "bruise" remedy. Anytime a blow, trauma, sprain/strain, or injury occurs to the body, this is the remedy of choice. I give my does a dose a couple times a day for two or three days, right after kidding. A dose is considered one or two of the pellets. You can purchase these at a health food store.

Homeopathic remedies come in different potencies; 6X, 30X, 6C, 9C, 30C. General use calls for a 30C potency, so look for the number on the container. Most stores have them displayed in a rack, as they are the size of a tube of chapstick. Because I am forever walking into or tripping over objects around the ranch, Arnica is a remedy I use often. In fact, it also comes in a gel that can be rubbed on the area affected, as long as the skin has not been broken.

Acidophilus

Lactobacillus acidophilus is a natural bacterium that resides in the digestive tract. It is essential to healthy processing of foods and nutrients. When the body is compromised by antibiotic usage, depletion of nutrients, stress (kidding, change of pens/pasture, new and/or different hay, traveling, etc.), these good bacteria may become depleted. Most breeders I know keep a tube of probiotics for just these occasions. I myself keep capsules of live acidophilus in the refrigerator. For me the easiest method is to give it to the goats in its original form or open the capsule and sprinkle on their grain. I also take a capsule a couple times a week to maintain the balance in my system.

Using alternative methods is an ongoing learning experience. It becomes a conscious awareness of trying natural products in conjunction with or instead of meds. I continually learn about new remedies that others have tried and found to be successful. Discussions and networking with other breeders results in a cumulative knowledge of using the best alternative available.

Until next time, enjoy your goats and kidding season, and share your alternative remedies with others!

From *Ruminations* # 55

Alternatives: Parasites
by Donna Geiser

As I sit here typing on the computer, I can look out my window and see the goats in the yard sunning themselves under the warm fall sunshine. It seems like only yesterday that the spring kidding frenzy was in full swing. Where did the time go?

Here at the ranch it is fall breeding season once again. Because we live in the arid high desert region of the Southwest, we do not have a big problem with parasites. We treat all the goats in the fall and again in the spring right after and during kidding season and this procedure works well for us. Parasite control seems to be the biggest problem facing breeders no matter where they are located. The subject is continually discussed on many goat lists, and I receive private e-mails regarding this matter.

When using injections or paste worming medications, we have to remember that a product that works in one area often may not work in another. Parasites that thrive in wet damp conditions will respond better to a particular product for that location. The same product may not necessarily work elsewhere.

If goats are in pasture conditions, one way to slow down the infestation cycle is to move the goats to a dry lot. This is an area where vegetation is sparse and can be easily cleaned. Administer the product and allow them to stay there for at least 24 to 36 hours. This process will minimize exposure to re-infestation in the pasture.

Constant overuse of commercial worming products has many breeders turning to an alternative: formulated herbal wormers. These are found on web sites, at feed stores and at animal supply companies. Some are packaged as pellets and some are loose; both are used as a top dressing on hay or grain rations. No withdrawal time is required for milk or meat consumption and they can be safely used on pregnant does.

What types of herbs are used in these products? I suggest making sure they come with a list of ingredients. Don't take anyone's word for what they contain. The following herbs are the most useful for controlling parasites no matter where you live:

Artemisia

This is the number one ingredient any herbal worming product should have. It is most well known for purging the body of parasites and is generally used in combination with other herbs. (Wormwood, southern wormwood, mugwort, and western sagebrush are in the Artemisia family.)

Ginger Root and Clove

Both calm the stomach and intestinal tract.

Mullein

This is a wonderful herb that controls coughs and supports the lungs.

Elecampane

This is another exceptional herb that supports the digestive and respiratory systems as well as being a natural expellant. Unfortunately, it isn't used in combination herbal wormers as much as it should be.

Acidophilus

This probiotic puts good bacteria into the stomach and supports the intestinal system. It should be given after a round of antibiotics. If you use a probiotics paste, you are using acidophilus.

Many herbal combinations can be found or created to control parasites. If we as proponents of alternatives assist the companies to improve their products, our animals will benefit. If you are undecided about whether the herbal wormers work, I suggest that you use a test group in your own herd. Investigate and examine the process that works best for you in controlling parasites. You may be pleasantly surprised!

From *Ruminations* #55

Oregano Essential Oil for Chronic Coccidiosis

A study in Italy used oregano essential to treat goats that were chronically infected with coccidiosis. They were given the essential oil daily for 30 days, and fecal samples were done on days 0, 10, 20 and 30. The fecal oocyst number started to decrease on the 10th day and had dramatically decreased by the 20th day. The species of oregano used was the Mediterranean variety, *Origanum vulgare L.*

(**Note:** Oregano essential oil should never be given undiluted.)

Goat Massage 101

by Terri Smith, RN

Even though many of us don't think about giving massages to our goats, doing so can provide many of the same benefits for them as for humans. Massage facilitates faster healing after an injury or surgery by stimulating the immune system and improving blood and lymph circulation. It helps restore flexibility to an injured limb, to aid in the prevention of future injury or debilitation. Strained or injured muscles heal faster when massage is applied, and fibrosis or scarring between adjacent muscle fibers can be prevented or broken down by repeated treatments.

Massage can calm and relax a nervous animal, making capture and handling for milking, veterinary care and transportation easier. A goat that has been massaged is more able to handle the stresses of showing, transport and breeding season.

Massage has the added benefit of being a good diagnostic tool for finding tumors, abscesses or other abnormalities. Massage is also an effective, noninvasive remedy for bloat. During the birthing process, massage may increase the release of beta-endorphins providing relaxation and natural pain relief.

Anyone can apply simple massage with a little basic information. Different methods and pressures can be used depending on the reason for the massage. A fast and strong massage rate will stimulate the goat, while a slow rate and constant rhythm may be sedating.

A basic understanding of caprine anatomy is important. While no one has yet written a book on goat massage, many good books and web sites on canine massage can provide you with a good overall idea about the musculature of your animals. Be careful around bones, as these attachments can often be sensitive. Use your fingertips or thumbs to work these areas and increase your pressure slowly. Keep treatments to 30 minutes or less.

Just as some goats are friendlier and like being touched more than others, certain goats will be more tolerant of massage than others. You may have to try several times before you are successful.

At the beginning of any treatment, start out with light, slow strokes, similar to petting, along the muscle in direction of the coat to calm the goat and acclimate it to your touch. Passive touch, or holding your hands still on the goat for a short time, can also be a calming way to start the session. Use firm but gentle strokes when massaging, gradually deepening your pressure as the tissues relax. Remember when applying massage that you are working the muscle bodies and their attachments to the bone.

Bloat

In addition to other treatments to relieve bloat, and depending on its severity, massage may be of some value. To treat bloat with massage, stand or kneel to the left of the animal and, using the fingertips and palms of your hands, massage both flanks, especially the left side (rumen), in a firm, circular, kneading fashion until the goat begins to burp or pass gas. As is the key with all massage, too little pressure is always better than too much. Pay attention to how the goat is responding. Feel for any masses or tumors while massaging the abdomen. This will become easier with practice as your fingers become more sensitive.

Leg Injury

For a leg injury, begin with calming strokes, and then gently massage the muscles around the shoulders for a front leg or around the haunches for a back leg. Using a kneading motion, thoroughly massage the muscles, working your way slowly down the front of the leg until you reach the hoof. Then massage back up the inside of the leg. Don't forget to work around the bones where the muscles attach. Repeat this 2–3 times using slightly deeper pressure each time. If the goat won't tolerate this, come back a short time later and repeat the session again. When working on an injury, giving several short sessions daily is more beneficial than one long session every few days. Do not massage a goat with a fever or around an infected area or on a limb with a circulatory disorder, as this could cause the infection to spread or blood clots to develop.

Calming Massage

Massage to calm a goat should be done in a slow and relaxed way. Always begin with calming strokes and be sensitive to the animal's reaction. As you probably know if you have worked with a variety of goats, developing a trusting relationship may take time and patience. Try rubbing the temples at the outside corners of the eyes. This is very soothing to an animal. Some goats love to have the area around their ears gently massaged.

Using your thumbs, find the long muscles just lateral to the spine and work down the back using a small, circular motion. Don't lift your hands as you move down the back; gently slide your thumbs to the new position. Massage around the top of the tail and the gluteal muscles. Repeat this 2–3 times. Massaging around the abdomen as described above is also soothing. Finish with more calming strokes.

Like humans, each goat is different. Each one will require a slightly different touch and some may never accept massage. If you start young with them, you will raise animals that are easier to handle at shows, easier to milk and more accepting of your touch when you need to give shots or work on an injury. Regardless of age, any goat in your herd can benefit from massage.

From *Ruminations* #55

Sources:

Constantinescu, G.M. 2001. *Guide to Regional Ruminant Anatomy Based on the Dissection of the Goat.* Wiley Blackwell.

Hourdebaight, J. 2003. *Canine Massage: A Complete Reference Manual.* Dogwise Publishing.

Letting Go

One of the best parts of raising goats is greeting the new kids that are born each year. Unfortunately, nature balances this new life with death, and we all will be faced with the death of a goat, at some point. *So, how do you know when to let go of your goat?*

With the exception of cases where an emergency requires the immediate action, the goatkeeper will have time to evaluate whether or not a goat should be euthanized or treatment discontinued. A number of factors will enter into such a decision. For some goat keepers, the decision to euthanize a goat may be purely economic: They can't afford to pay for the treatement or don't want to spend the money on the goat.

Others, perhaps having seen other goats suffer from a disease such as *mycoplasma* pneumonia or CAEV, or learning from the vet that a disease is terminal, want to spare the goat from a long, downhill course.

In other cases, people may not have the accommodations, time or money to provide the care that a dying goat needs. Whatever the situation, knowing when to let go of a goat is a personal decision.

Often pneumonia, enterotoxemia, parasitism or some other malady will kill a goat in short order. This can be hard, especially when it is unexpected and, particularly, when a beloved goat is the one that dies. Other times, a goat will develop a chronic illness and you will have to make decisions about whether to continue treatment, how much diagnostic work and treatment to have done, and whether it's time to provide only palliative care or have the animal euthanized.

Carnation, a six-year-old Nigerian dwarf who was positive for CAEV, had never shown any signs of the disease or had any health problems. Then gradually she began to eat less and to lose weight. She ultimately developed a cough and then I noticed that she was being increasingly bullied by her niece. I isolated her, discovered that she had a fever, wormed her because she looked anemic, and began treatment with Penicillin.

After a week of treatment, she showed no improvement. I had started giving her roughage (blackberry shoots, salal and fir branches) and hand-feeding her, as well as encouraging her to drink hot water and Goat Magic. I also made her a jacket out of a fleece blanket to wear at night for extra warmth.

She began to grind her teeth, signalling that she was in pain, so I added Banamine to the regimen. I then started her on Bio-Mycin, which she received for a week. Still no improvement; in fact, she became quite wobbly, was eating less and coughing more. I now had to make a decision about whether her quality of life had decreased so much that it was time to let her go.

Alice Villalobos, DVM, has developed a Quality of Life Scale, called HHHHHMM, which is meant to be used by pet owners to determine when an animal should be let go. The factors to consider include Hurt, Hunger, Hydration, Hygiene, Happiness, Mobility and More good days than bad days. Each of these categories is scored from 1 (poor) to 10 (good). A score of 35 or less indicates an unacceptable quality of life. This scale is useful for determining whether a goat's quality of life is acceptable or not, as well.

Hurt

Does the animal have adequate pain control? Goats that are in pain will often grind their teeth and withdraw from other goats. You may be able to control pain with ibuprofen, aspirin, Banamine or other nonsteroidal anti-inflammatory drugs. In some cases dexamethasone or another steroid may be used. All of these drugs can have side effects that will limit the length of time they can be used. Uncontrolled pain is a good reason to euthanize a goat.

Hunger

Many times sick animals will lose their appetite; it's a normal part of the dying process. Is the goat eating enough, or can it be fed by hand or get calories through a feeding tube? As in the case of Carnation, some sick goats can be supplemented with roughage, fruit or other foods that they like to keep them going. These things, along with probiotics and even some herbs can help the goat maintain a healthy digestive system and stimulate the appetite.

Hydration

When a goat completely loses interest in drinking and becomes dehydrated, the situation is dire. Can the goat be encouraged with flavored hot water, electrolytes, or a home remedy like Goat Magic? In some cases, subcutaneous injections of sterile water can keep an animal going long enough to improve. If finances, skill level or other factors allow it, a goat can be given an IV. If a goat isn't interested in eating or drinking at all, she may be at the end of her life.

Hygiene

In the case of goats, hygiene most likely will relate to whether the animal can get up and move around, or whether it has diarrhea and is unable to stay clean. In some situations, goat keepers can help with this by helping the animal move and cleaning it up. One question to ask is whether, even with help, the animal will improve.

Happiness

Goats express happiness by their responsiveness, playfulness and lack of depression. Being isolated, in itself, will make a goat—a herd animal—less happy. Does the goat make eye contact? Does she respond to being petted, brushed or even massaged? Does he seem glad to see you?

Mobility

Can the goat get up and walk around? Is he having seizures or a hard time navigating well? Some enterprising goat owners have even built carts for animals that were unable to get around due to an injury, and found that this improves their quality of life.

More good days than bad

The first six criteria will factor into this one. Is the bond between goat keeper and animal still there? Is she suffering? Is he improving, staying the same or getting worse? Or do the poor scores on all the factors that go into quality of life equal a decision to let her go?

I had instinctively considered all of these factors in my decision-making about Carnation and decided that she was dying and needed to be kept comfortable. But then, after two weeks of no treatment other than pain medication, hand-feeding and watering, I concluded that she wasn't ready to die. I called the vet, who prescribed dexamethasone, a steroid, and Naxcel, another antibiotic, and I began treatment again. I also added to her comfort measures an electric warmer to sleep on during the cold nights.

Within an hour of the dexamethasone injection, she showed drastic improvement. Then, during the ten days of Naxcel, she began to go downhill again, never losing her cough and becoming unsteady on her feet. I re-evaluated the situation, asked for some help on a goat list, and decided to try the steroids one more time, in the interest of keeping her comfortable.

At this point, it was clear to me that she was dying, and it was time to focus on keeping her comfortable and considering euthanasia when the balance inevitably shifted to an unacceptable quality of life. Carnation died one week later, at her own chosen time.

Source:

Villalobos, A.E. 2006. Canine and Feline Geriatric Oncology Honoring the Human-Animal Bond, Blackwell Publishing, Table 10.1.

Euthanasia

As a caring goat owner, you will want your goat to have as peaceful and painless a death as possible. If you can afford it and have the access, you can either have a veterinarian come to your farm or you can take the goat to his or her office, if it is in good enough condition. In some cases, of course, the goat will already be at the animal hospital or veterinary school if it is being treated for a serious illness.

You will also have to decide whether or not you want, or can, be there when the veterinarian puts your animal down. This is a personal decision, and each case may be different.

The most common method for euthanizing an animal is lethal injection with a barbiturate, which will shut down the animal's nervous system. The American Veterinary Medical Association (AVMA) Panel on Euthanasia recommends the use of pentobarbital for this purpose. Normally the animal's heart will simply stop beating and it will stop breathing. Other than some natural deaths, lethal injection is probably the best way for an animal to pass on. But not everyone has access to such drugs or a professional with the expertise to use them.

If, like many people, you end your suffering goat's life with a firearm, make sure to be respectful and to remove the animal from the herd to a safe, quiet place when possible. Goats can become alarmed, and do not need the stress of a gunshot.

While many people don't like the idea of shooting an animal, if done properly, it will cause immediate unconsciousness. And it may be the only option available to put an animal out of its misery.

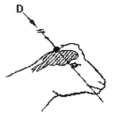

A .22-caliber long rifle, 9 mm or .38-caliber gun can be used by an individual who is trained in the use of guns. Hold the muzzle of the gun at least 4–10 inches from the skull and shoot just behind where the horns would be, and toward the back of the chin. Kids less than four months old may be shot from the front of the mid-forehead, just above the eyes and aiming along the angle of the neck.

Make sure that everyone in the area is out of the line of fire and that you are in compliance with any state and local gun laws.

Other methods of humane euthanasia may be available, depending on the individual and farm circumstances. (See Resources for further information.)

Disposal of the Body

Over the years, I have experienced the deaths of many goats. How I disposed of their bodies has varied widely. State laws regulate how dead livestock should be disposed of, so that's the first place to look if you are unsure of what is permissible or legal. In addition, some federal laws govern this area, in order to protect endangered species and other wild animals.

One thing that is important to the health of the goat herd is to remove the animal's body from the barn or other area that is accessible to goats or other animals as soon as possible to avoid spreading disease.

If you have ended your goat's life with a gunshot, be aware that bullets and shot contain lead, which is known to poison some birds.

Goats that have been euthanized with a barbiturate or other lethal drug are left with a toxin in the body. This is a poisoning hazard to eagles, crows, cougars, dogs, vultures and any other animal that eats carrion. NEVER leave the body of an animal whose life was ended this way in a field or the woods with the intention of giving wild animals something to eat.

Some goat diseases are transmissible to other animals, particularly deer. Regardless of the method used to end a goat's life, the body should be properly disposed of to avoid spreading disease.

Burial

Burial is the most common and least expensive method for disposing of a dead goat. If state and local laws allow it, burial may be the best option for most goat keepers.

In parts of the country where the ground freezes in the winter, or where owners who don't have enough property or have rocky or clay soil that is impossible to dig without equipment, burial may not be feasible. In some cases, a neighbor or friend may be willing to bury the animal on their property, or you can find a pet cemetery, if the goat is special.

If you do bury an animal, make sure that it is buried deeply enough (4–8 feet) and, if necessary, is covered with rocks or other objects that will discourage digging by dogs or other animals. State law may govern maximum and minimum depth.

Never bury the animal near a well, a creek or other water source, or in wetlands or a floodplain.

Composting

A fairly recent trend in livestock disposal is composting. A fairly simple box can be built in which to compost dead goats or some of the principles can even be used when burying a goat.

To simplify: Place a layer of sawdust (12–15 inches) in a bin, put the goat on top on its back and cut open the chest and abdominal cavities, as well as the organs. Cover with another 12–15 inches. Dampen with water as needed, for a 50–60% moisture content. It should look damp and be moist to the touch. More than one goat can be composted, in layers. The whole thing should be covered with a tarp to keep rain and animals out. (See Resources for further information.)

Cremation

When my first goat, Jinx, died I was so heartbroken that cremation seemed the only option. I still have her ashes on my mantle. (Urns for ashes are available online in a range of prices and styles or you can use special container of your choosing.)

Often funeral homes that cater to people will do pet cremation. In some cities, businesses dedicated solely to animal cremation can be found. In some cases, the cremation is communal; others offer private cremation of individual animals. To find a service in your area, contact the nonprofit International Association of Pet Cemeteries and Crematoriums (www.iaopc.com/).

One of the disadvantages of cremation is that, without a necropsy, you will never learn what caused the goat's death and what other problems may have been in play (something I regretted with Jinx). In the case of a beloved goat that died of old age, though, it may be a very reasonable option.

Performing a Necropsy

While not a method of disposal, having a necropsy performed by a veterinarian or veterinary school will serve the dual purpose of obtaining information about the cause of death and having the goat's body disposed of.

A necropsy is an animal autopsy. Ideally, it should be done as soon as possible after the death. If you are unable to get the goat to a veterinary school or a veterinarian right away, the goat should be refrigerated, if possible.

Some goat keepers do their own necropsies, in cases where the option to have a necropsy is not available (e.g., no access to a veterinarian). (See Resources for information on performing your own necropsy.) If you are brave enough to attempt a necropsy on the farm, you will still have to find a way to dispose of the body promptly after it is completed.

A goat should be professionally necropsied when multiple, unexplained deaths have occurred in the herd, when multiple goats have aborted, when a goat dies suddenly or when you simply want to learn what killed the goat. In some cases a veterinary school will accept an animal for donation, either prior to euthanasia or after, for teaching purposes.

Routinely having a necropsy done when an animal dies has several advantages. It allows you to monitor the herd to learn what disease issues you are facing. Some of these diseases may be hidden and not even contribute to the death, yet still have a herd impact.

If you also request testing for levels of copper, selenium and other minerals, you can receive valuable information on how well your feeding plan is working or where it can be improved. You also may learn of parasite problems that weren't apparent.

An added benefit of having a necropsy done by a professional is that the veterinarian or veterinary school will also dispose of the body.

Other Methods of Disposal

In some areas, the local landfill or waste facility will accept animals under a certain size. In other areas, goat keepers don't have that option but may be have a rendering company that will come and pick up the animal.

How you dispose of the body of a goat that has died on your farm or in your home will depend on numerous factors and may vary from goat to goat. It's worthwhile to consider in advance how you will handle this issue before you ever have to face it, particularly if you have a close relationship with that goat. This avoids making decisions that you may later regret, because you were operating out of grief.

Bereavement

Although everyone has a different grief process, many of us go through a range of emotions when one of our animals is dying. These may include some or all of the stages of dying that Elizabeth Kubler-Ross first enumerated. These are denial, anger, bargaining, depression and acceptance. Along with these emotions, we may feel guilt, as well—for not recognizing the illness soon enough, for not acting quickly enough or for not doing enough. The acceptance phase may be necessary to finally acknowledge that the goat's death is inevitable and/or to have it euthanized.

Even after the death, feelings of grief may continue. This may be particularly true if the goat had a long dying process that strengthened the animal-human bond. Understanding that this is normal can help in dealing with these feelings and eventually getting over or at least accepting the loss.

Some people may not understand the bonding that can occur between a human and a goat, and discount your feelings. This is the time to seek out other goat people and share your experiences. You can find a sympathetic ear on some online goat lists, such as Yahoo groups; many of the members will have had a similar experience with a beloved goat.

Memorials

Another way to deal with sorrow over the loss of a goat is a memorial. This can be a garden stone, an urn for the ashes of a goat that was cremated, or posting the goat's picture and a few words on a web site, either your own or one of the "Rainbow Bridge" sites. Another option for breeders would be to run a farm ad celebrating the goat—particularly if it was a champion or prize milker.

Counseling

Some veterinary offices also have grief counseling services that are geared to the loss of companion animals. Others can suggest other options, such as support groups, grief counselors or hotlines.

Goat Therapy

Finally, one of the best ways to get over the loss of a loved goat is to hang out with the goats that you still have. Nothing can brighten the spirit or pull you out of the doldrums like a goat!

Resources

Understanding Your Goat

Goat Terminology
www.goatworld.com/articles/terminology.shtml

Medical Assessment Form
www.cometothefarm.com/medical_assessment.htm
To assist in working with a vet to diagnose medical problems

Animals in Translation: Using the Mysteries of Autism to Decode Animal Behavior, by Temple Grandin and Catherine Johnson. 2006. Orlando, Florida: Harcourt. 359 p. $15.00.

Routine Goat Care

Fencing and Housing
Wisconsin Dairy Goat Association
www.wdga.org/widairygoatassoc/resources+for+farmers/goat+housing.asp
A variety of links to dairy goat housing

Colorado State University
http://cerc.colostate.edu/Blueprints/Goat.htm
Goat Equipment and Housing Plans

Langston University Goat Research
www.luresext.edu/GOATS/library/fact_sheets/g02.htm
Housing Your Goat

Maryland Small Ruminant Housing and Equipment
www.sheepandgoat.com/housing.html

Feed
Nutrient Requirements for Small Ruminants, by National Academies Press, $103.20
Online at http://books.nap.edu/openbook.php?record_id=11654

Evaluating Goat Feeding Management through Body Condition Scoring
http://smallfarms.wsu.edu/animals/EvalGoatFeeding.html

An Introduction to Feeding Small Ruminants
www.sheepandgoat.com/articles/feedingsmallruminants.html

Nutrient Requirement Calculators
www2.luresext.edu/GOATS/research/nutr_calc.htm

Diet for Wethers: A Guide to Feeding Your Wether for Health and Longevity, by Carolyn Eddy. $14.95

Pasture Management
Maryland Small Ruminant page
www.sheepandgoat.com/pasture.html
Lots of pasture management links

Parasite Control
The Parasite
www.apacapacas.com/parasites/
Images of parasites

Southern Consortium for Small Ruminant Parasite Control
www.scsrpc.org
The latest in goat parasite control

The Secret Lives of Parasites
www.goatbiology.com/parasites.html
Includes a good drawing of the life cycle of the liver fluke

Comparison and info on various lab diagnostic techniques
http://cal.vet.upenn.edu/projects/dxendopar/index.html#fecal

Integrated Parasite Management for Livestock (1999)
National Sustainable Agriculture Information Service
http://attra.ncat.org/attra-pub/PDF/livestock-ipm.pdf

Kid Care

The Wimpy Kid Page
www.goatwisdom.com/ch1baby_care/wimpykid.html
Lots of information on dealing with a weak newborn kid

Kid Care and Information
Fiasco Farm
http://fiascofarm.com/goats/kid-care.htm
Week by week information on kid care

Tube Feeding Weak Kids
www.boergoats.com/clean/articleads.php?art=39
A novel method for tube feeding, to prevent drowning the kid

Creep Feeding Kid Goats
www.boergoats.com/clean/articleads.php?art=650
Good discussion of creep feeding

Building a Disbudding Box
http://kinne.net/disbbox.htm

Smoke Gets in Your Eyes
http://kinne.net/disbud.htm
A good article on disbudding by Maxine Kinne

Why Horns?
www.goat-idgr.com/Default.aspx?tabid=97
An alternative view on goat horns

Notes on the Care of Miniature Goat Kids
www.goat-idgr.com/Default.aspx?tabid=100
Bottle-feeding and other info geared to minis

Breeding, Pregnancy and Kidding

A Beginners Guide to Breeding Your Goat (2005)
Goat Source
www.goatsource.com/Breeding Your Goat.pdf

Excel spreadsheet to track breeding and kidding
http://fiascofarm.com/goats/breeding-kidding_sheet.html

Breeding/Kidding Due Date Calculators
www.boergoats.com/tools/gestation.php
www.goatworld.com/articles/pregnancy/pregnancy.shtml

Causes of Infectious Abortion in Goats (2006)
Alabama Cooperative Extension Service (ACES)
www.aces.edu/pubs/docs/U/UNP-0079/

Birthing Kit Contents
http://fiascofarm.com/goats/birthingkit.html

Health Issues

General Veterinary and Disease Information
International Veterinary Information Service (IVIS)
www.ivis.org
A convenient way to access information on veterinary medicine. Free to clinicians, researchers, librarians, educators and veterinary students with the financial support of private, public and corporate sponsors. Registration is required.

Merck Veterinary Manual Online
www.merckvetmanual.com/

Fact sheets on a variety of animal diseases from Iowa State University
www.cfsph.iastate.edu/DiseaseInfo/factsheets.htm

Bloat
Detailed article on bloat in goats
http://kinne.net/bloat.htm

Brucellosis
Frequently asked questions on brucellosis
Centers for Disease Control (CDC)
www.cdc.gov/ncidod/dbmd/diseaseinfo/brucellosis_g.htm

Caprine Arthritis Encephalitis Virus (CAEV)
Book chapter on CAEV
www.ivis.org/advances/Disease_Tempesta/bertoni/chapter.asp?LA=1

Fact sheet on CAEV
www.ivis.org/advances/Disease_Factsheets/caprine_arthritis_encephalitis.pdf

Update on Caprine Arthritis Encephalitis
Washington State University College of Veterinary Medicine
www.vetmed.wsu.edu/depts_WADDL/caefaq.aspx

2007 Fact Sheet on CAEV
University of Iowa
www.cfsph.iastate.edu/Factsheets/pdfs/caprine_arthritis_encephalitis.pdf

Caprine Arthritis-Encephalitis Virus
www.vet.uga.edu/vpp/clerk/logan/
A good article on CAE

Caseous Lymphadenitis (CLA)
PHL Associates
www.phlassociates.com/
Custom CLA vaccines for large goat herds

Copper Deficiency
U-Say Ranch
www.u-sayranch.com/goats/copper.html
Step-by-step information on treating copper deficiency

Listeriosis
Goat Polio or Listeriosis?
www.tennesseemeatgoats.com/articles2/listeriosis.html
An article comparing polioencephalomalacia with listeriosis, with a dosage chart for procaine penicillin used to treat listeriosis

Johne's Disease
www.johnes.org/goats/
Factual information, true stories of Johne's disease in goat herds, epidemiology, graphics and more

http://ohioline.osu.edu/vme-fact/0003.html
Ohio State University Fact Sheet on Johnes

Mastitis
Gangrene Mastitis Pictures
Black Mesa Ranch
www.blackmesaranch.com/animals/goat_lineage/goats_lineage_42nougatb.htm

Mastitis in Dairy Goats
University of Florida, IFAS Extension
http://edis.ifas.ufl.edu/pdffiles/DS/DS12000.pdf

Mycoplasma
Caldwell, Gianaclis. 2008. Mycoplasma: Knowledge is Power to Fight this Deadly Goat Disease. *Dairy Goat Journal*
www.dairygoatjournal.com/issues/86/86-6/mycoplasma.html
Excellent article on mycoplasma

Poisoning
Plants Poisonous to Livestock, Angela McKenzie-Jakes, Extension Animal Science Specialist Florida A&M University Research and Extension Programs
www.famu.edu/goats/UserFiles/File/Poisonous_Plants_to_Livestock_Website(1).doc

IVIS
www.ivis.org
Extensive information on poisonous plants

Q Fever
An article on Q Fever
www.cdc.gov/healthypets/diseases/qfever.htm

Hatchette, et al. 2000. Goat-Associated Q Fever: A New Disease in Newfoundland
www.cdc.gov/ncidod/eid/vol7no3/pdfs/hachette.pdf

Hatchette, et al. 2003. Natural History of Q Fever in Goats
www.liebertonline.com/doi/abs/10.1089/153036603765627415?cookieSet=1&journalCode=vbz

Q Fever in Goats, Sheep, and People
Wyoming State Veterinary Laboratory
http://wyovet.uwyo.edu/Diseases/2004/Qfever.pdf

Rabies
Detailed information on rabies
CDC
www.cdc.gov/RABIES

Scrapie
National Scrapie Eradication Program
Videos, fact sheets and other info on scrapie eradication
www.aphis.usda.gov/animal_health/animal_diseases/scrapie/

Hourigan, et al. 1979. Epidemiology of Scrapie in the United States.
www.bseinquiry.gov.uk/files/mb/m08b/tab64.pdf

Cornell University. Genetics of Scrapie in Sheep www.ansci.cornell.edu/sheep/management/health/scrapiegenetics.htm

Urinary Stones
Kinne, M. Urolithiasis in Pygmy Goats
kinne.net/urincalc.htm

Urolithiasis in Small Ruminants: Surgical and Dietary Management
www.aasrp.org/hot_topics/2004/August%202004/Urolithiasis/Cornell%20Urolith%20Surgery%2004.doc

Medications

Haskell, Scott R.R., and Theresa A. Antilla. 2001. *Small Ruminant Clinical Diagnosis and Therapy*
www.rmncsba.org/SMALLRUMINANT.pdf

Food Animal Residue Avoidance Databank (FARAD)
www.farad.org
Lists prohibited drugs, regulatory actions and other information on avoiding drug residues in food animals.

Conditions for Producers' Use of Livestock Drugs
Washington State University
www.vetmed.wsu.edu/courses-jmgay/VMADProducerDrugs.htm
Geared to veterinary students, this guide provides information on the legal requirements for livestock producers to use drugs in food-producing animals

Goats: Medicine and Surgery
www.filemanage.co.uk/pubs/Farm_pets.pdf

Database of Approved Animal Drug Products
www.accessdata.fda.gov/scripts/AnimalDrugsAtFDA/

FDA's Center for Veterinary Medicine
www.fda.gov/cvm/default.html

Vaccination Protocol for a Goat Herd
Alabama A&M and Auburn Universities
www.aces.edu/pubs/docs/U/UNP-0090/UNP-0090.pdf

Natural Care and Home Remedies

Lans, C., et al. 2007. Ethnoveterinary medicines used for ruminants in British Columbia, Canada. *Journal of Ethnobiology and Ethnomedicine* 3: 11. Available online at www.ethnobiomed.com/content/3/1/11

Herbal Medicine: The Natural Way to Get Well and Stay Well, by Dian Dincin Buchman. New York: Gramercy. 310 p. Out of print.

The Complete Herbal Handbook for Farm and Stable, by Juliette de Bairacli Levy. 1991. Great Britain: Mackays of Chatham. 471 p.

Goats: Homeopathic Remedies, by George McLeod. 2004. UK: Random House. 192 p.

Herbal Recipes for Farm Animals: Chemical Free Alternatives, by Diana L. Manseau. 1997.

Natural Goat Care, by Pat Coleby. 2001. Austin, Texas: Acres USA. 371 p.

Mosby's Handbook of Herbs & Natural Supplements, 3rd ed., by Linda Skidmore-Roth. 2006. St. Louis, Missouri: Elsevier Mosby.

Letting Go

Euthanasia
The Emergency Euthanasia of Sheep and Goats
University of California at Davis
www.vetmed.ucdavis.edu/vetext/INF-AN/INF-AN_EMERGEUTHSHEEPGOAT.HTML

On Farm Euthanasia of Sheep and Goats
Ontario Ministry of Agriculture, Food and Rural Affairs www.omafra.gov.on.ca/english/livestock/animalcare/facts/info_euthanasia_shgt.htm

Composting
Article on disposing of dead goats from the Maryland Cooperative Extension
www.sheepandgoat.com/articles/compost.html

Composting Dead Animals Fact Sheet
http://crawford.extension.psu.edu/Agriculture/Composting.htm

Necropsy
If you live in an area without access to a veterinarian and feel confident enough to do your own necropsy, go to http://books.google.com and type in the book name "The Laboratory Small Ruminant." That will take you to some online content that gives an overall description of how a basic necropsy is performed.

A goat necropsy, with photos (graphic)
www.geocities.com/amyk1111/goat.html

Bereavement
A web site for pet lovers grieving over the death of a pet or an ill pet. Includes personal support, thoughtful advice, The Monday Pet Loss Candle Ceremony, tribute pages and more.
www.petloss.com/

A virtual home for your departed goat or other animal
www.rainbowbridge.com

Other Goat Care Information and Supplies
Books
Diseases of the Goat, 2nd ed., by John Matthews. 1999. Oxford: Blackwell Science. 364 p. Out of print.

Goat Health Handbook: A Field Guide for Producers with Limited Veterinary Resources, by Thomas R. Thedford. 1983. Morriston, Arkansas: Winrock International. 123 p. Out of print.
This book is essential for those caring for goats in areas where vets and medications are not available.

Goat Husbandry, by David MacKenzie. London: Faber and Faber. 1970. 366 p. Out of print.
From Britain, this is still a good reference book even though it is dated.

Goat Medicine, by Mary C. Smith and David M. Sherman. 1994. Baltimore: Lippincott, Williams and Wilkins. 620 p., $80.00.
Although somewhat dated, this is the most complete medical text available. Much of *Goat Medicine* is available online at http://books.google.com

Guide to Regional Ruminant Anatomy Based On the Dissection of the Goat, by Gheorghe M. Constantinescu. 2001. Hoboken, New Jersey: Wiley Blackwell. 252 p.

National Goat Handbook. 1992. USDA.
http://outlands.tripod.com/farm/national_goat_handbook.pdf

The Merck Veterinary Manual, 8th ed., Susan A. Aiello, ed. 1998. Whitehouse Station, New Jersey: Merck & Co. 2305 p.
Also available online at http://merckvetmanual.com

The Goatkeeper's Veterinary Book, 3rd ed., by Peter Dunn. 2004. UK: Old Pond Publishing. 227 p.

Discussion Lists
http://goatconnection.com/discus/
Online forum on various goat topics

Washington State University goat discussion list
http://lists.wsu.edu/mailman/listinfo/goats
This e-mail list, made up of long-time goat keepers and veterinary professionals, can help with goat health care problem-solving

Yahoo discussion groups
http://groups.yahoo.com
Discussion groups on all aspects of goats and goat keeping

Laboratories
Washington Animal Diagnostic Disease Laboratory (WADDL)
Washington State University
Phone: 509-335-9696
www.vetmed.wsu.edu/depts_WADDL/

UC Davis Veterinary Genetics Laboratory
www.vgl.ucdavis.edu/
DNA Typing and other genetic tests

Periodicals
American Association of Small Ruminant Practitioners (AASRP)
PO Box 611
10220 Dixie Beeline Highway
Guthrie KY 42234
www.aasrp.org/
Goat owners can join as associate members and receive the quarterly *Wool and Wattles* newsletter, which has helpful info from goat vets and journal abstracts.

Dairy Goat Journal
145 Industrial Drive
Medford WI 54451
800-551-5691
www.dairygoatjournal.com

Ruminations Magazine
PO Box 859
Ashburnham MA 01430
Editor@smallfarmgoat.com
978-827-1305
www.smallfarmgoat.com

CDs
Small Ruminant Production Management and Medicine
www.infovets.com
Goat care CD-ROM, manual and online look-up, provide an easy reference for many goat health problems.

Supplies
Caprine Supply
www.caprinesupply.com
800-646-7736

Hamby Dairy Supply
www.hambydairysupply.com
800-306-8937

EZ-ID
877-330-EZID (3943)
Colorado: 970-351-7701
E-mail: EZID@avidid.com
www.ezidavid.com
Microchips, ear tags, scanners

Hoegger Goat Supply
800-221-4628
www.hoeggergoatsupply.com
Specializing in goat supplies and advice

Jeffers
www.jefferspet.com
800-533-3377

KV Vet
www.kvvet.com
800-423-8211

Light Livestock Equipment and Supply
866 999-AVA1(2821)
http://lightlivestockequipment.com
Fecal testing kits, flotation solution, microscopes and more

Livestock Concepts
http://livestockconcepts.com
800-225-7399
Heated mats for sick animals, and other products

Med/Supply Partners
www.medsupplypartners.com
Vacutainer tubes and other medical supplies

Wiggins & Associates
503 SW Victoria Ct
Gresham OR 97080
www.wigginsinc.com
800-600-0716

Supplements, Herbs and Capsules
Capsuline
www.capsuline.com/Frontend/Index.aspx
Capsules for copper or supplements

Mountain Rose Herbs
PO Box 50220
Eugene OR 97405
customerservice@mountainroseherbs.com
800-879-3337
www.mountainroseherbs.com
Organic herbs and teas

1-800 Homeopathy
www.1-800homeopathy.com
Informational articles and homeopathic potencies

Puritan's Pride
www.puritansale.com
Vitamins and supplements

Web Sites
Come to the Farm
http://cometothefarm.com
Resource info, classifieds and goat products

Fiasco Farm
www.fiascofarm.com
One of the most comprehensive and useful goat sites online!

Goat Connection
http://goatconnection.com
Their mission is to provide an opportunity for goat owners/breeders to find "everything goat" while providing an economical and effective way for companies and individuals to present their products to the world.

Goat World
www.goatworld.com

Goat Care Practices (booklet), 2000. UC Davis.
www.vetmed.ucdavis.edu/vetext/INF-GO_CarePrax2000.pdf

The Goat Dairy Library
www.goatdairylibrary.org/
A Wisconsin-based site with lots of dairy goat links and information, including a wide variety of forms for health and other management.

Maryland Small Ruminant Pages
www.sheepandgoat.com

New York State 4-H Dairy Goat Project
www.ansci.cornell.edu/4H/dairygoats/

Index

A

abomasum, 7–8, 123
abortion, 47, 63, 106, 116, 134
abscess, 100, 113–14, 169, 180, 193, 217
acidophilus, 214, 216
acidosis, 24, 26, 85, 186, 187, 190
albendazole, 79, 197
albon, 103, 189
allergy, 170
ammonium chloride, 166, 204
amoxicillin, 176, 177, 197
ampicillin, 168, 177
amprolium, 104
anaphylactic reaction, 41, 170
anemia, 22, 30, 104, 119, 120, 136, 142, 144, 175
anesthesia, 89, 90, 92, 96, 185
anthelmintic, 20, 30, 148, 197
anticoccidial, 103, 198
antitoxin, 90, 94, 131, 132, 163
apple cider vinegar, 26, 207
arnica, 214
artemesia, 209, 205
arthritis, 107, 108, 115, 138, 140, 176
aspirin, 92, 94, 148, 170, 182, 192, 199, 201, 202, 221
artificial insemination, 51–52
azalea, 15, 152
Azium, 182

B

Baerrman test, 144
baking soda, 23, 25, 85, 105, 132, 162, 200, 201, 202, 203, 204, 213
Banamine, 37, 132, 140, 148, 170, 180–81, 192, 220, 221
barber pole worm, 30–31, 142–43
Baytril, 177
B-Complex, 129, 130, 151, 178, 191
beet pulp, 23, 205
behavior, 1, 10–11, 13, 45, 69, 72, 96, 137, 157–58
Benadryl, 128, 190
bereavement, 227
Betadine, 28, 94, 98, 193
Bio-mycin, 129, 130, 172–73, 220
biosecurity, 173
Biosol, 189
birth defects, 48, 79, 188, 197, 210
bloat, 8, 23, 105, 131, 201, 217–18, 231
blood
  drawing, 40–41
  samples, 40, 50, 67, 153
blood stop powder, 193
bluetongue, 80
body condition, 14, 125, 135, 228
bolus, 22, 119–23, 189
Border disease, 64, 80
Bo-Se, 5, 47, 48, 81, 132, 178–179
BOSS, 24
Bovatec, 198
bottle-feeding, 89, 96, 230
bottle jaw, 30, 31, 142, 144
brassica, 80
breeding, 46, 47–48, 51–52, 53, 109, 179, 184, 202, 212, 215, 230
brucellosis, 40, 64, 106, 115, 231
Bryonia, 209
buck rag, 46
burdizzo, 94, 96
burial, 224
Bute, 170, 192

C

CDT, 42, 48, 32
CMPK, 54, 55
CMT, 139, 141, 195
Cache valley virus, 64
campylobacteriosis, 64
calcium, 22, 23, 26, 178, 186, 187
  deficiency, 53–60, 62, 182

gluconate, 187
calculi, urinary, 22, 43, 60, 94, 164–66
cambendazole, 48
caprine arthritis encephalitis virus, 40, 48, 81, 107–12, 139, 140, 145, 205, 220, 231
caseous lymphadenitis (CLA), 42, 113–14, 134, 147, 231
castration, 93–96, 105, 163, 165
ceftiofur sodium, 174
chlamydia, 47, 64, 115, 116, 117, 145, 172, 176, 194
chlamydiosis, 64, 115–17
chloramphenicol, 156, 175
chlorine, 22, 26
choking, 26, 83
Citatron, 180, 192
citronella, 208
Clostridium, 42, 114, 125, 131, 139, 149, 163
clove, 216
CMPK, 54–55, 187, 191
coccidia, 33, 101, 144, 188, 189, 198, 205, 209
coccidiosis, 101–04
colostrum, 53, 81, 82, 109,110, 112, 134, 200
comfrey, 66, 205, 206
copper, 21, 22, 34, 133, 226, 231, 238
  deficiency, 57, 79, 118–23, 124-26, 160
  poisoning, 21
Corid, 104, 149, 198
cottonseed, 17
cremation, 225
Cydectin, 197

D

dandelion, 209
dandruff, 25
Deccox, 47, 103, 198
Dectomax, 197
deformity, 80
dehydration, checking for, 5
deworming, 30–31, 47, 144,
dexamethasone, 138, 146, 151, 181–83, 184, 221, 222
diarrhea, 5, 32, 64, 65, 85, 101– 102, 104, 124, 131– 32, 134– 35, 141, 175, 181, 191, 204
Dichelobacter nodosus, 133
Di-Methox, 103–04, 176, 189
disbudding, 88–92
disposal, body, 224–26
Dopram, 185
dosage conversion, 171
doramectin, 197
drenching, 170–71
Drenchrite test, 143

E

E. coli, 104, 139, 140, 172, 180, 184, 189, 193
elastrator, 95, 96
elecampane, 216
electrolytes, 83, 84, 187, 190, 221
emasculator, 94
encephalitis, 108, 109, 138
enrofloxacin, 177
enterotoxemia, 85, 104, 131–32, 150, 163, 220
epinephrine, 41, 180
eprinomectin, 197
Eprinex, 197
esophageal groove, 7
ethology, 10
euthanasia, 223
erythromycin, 176
Excenel, 174
eyebright, 207

F

FAMACHA, 30–31, 144
FARAD, 168, 233
FastTrack, 191
FDA, 168
fecal exam, 32, 33–36
feeding, 14, 165
  kids, 87
  tube, 82, 83–84, 229
fenbendazole, 143, 197
fencing, 13, 19, 127
fever, milk, 22, 60, 62, 201

fighting, 48, 69, 127, 129
fir, 203–205
flies, 102, 145, 201, 208
floppy kid syndrome, 85
florfenicol, 174–75
fluke, liver, 30, 142, 144, 229
flunixin meglumine, 180–81
foot rot, 27, 29, 133, 172, 174
freemartin, 79
fungus, 159

G

gentamicin, 177
gestation,
　calculator, 230
　length of, 2, 46
ginger, 202, 203–04, 205, 216
　beer, 201
Goat Magic, 200, 221
goiter, 25
　milk, 100
　plants causing, 80
gossypol, 117
grain, 15, 43

H

haemonchus contortus, 30, 34, 142
hairy shaker disease, 80
hawthorn berries, 213
hay, 14, 15
heart rate, 4, 54–55
　in polio, 150
　rapid, 80, 144, 152
hermaphrodite, 78
honey, in wound healing, 201
hoof trimming, 27–29, 193
housing, 13, 228
　for bottle-fed kids, 82
　for young bucks,
hydration, 14, 163, 221
hypocalcemia, 53–60
hypokalemia, 85

I

inbreeding, 46, 49
injections
　giving, 37–39
　lethal, 223
　painful, 169, 172, 175, 180
insemination, artificial, 51, 52
iodine, 7, 22, 24–25, 73, 75, 80, 81, 87, 98,
　100, 162, 169, 193
　deficiency, 80
ivermectin, 137, 143, 191, 197
Ivomec, 197

J

Johne's disease, 119, 134–35

K

kaolin pectin, 105, 191
kelp, 24, 25
keratoconjunctivitis, 115, 145–46
ketosis, 61–62, 180
kidding, 45, 69–78, 206, 214–15, 230
　and CAEV prevention, 48, 111
Koppertox, 133

L

LA-200, 64, 67, 116, 145, 172–73
Labor, 68, 69–73, 202
　augmenting with oxytocin, 184
　induction, 180, 182, 183, 184
　premature, 183
lactated ringers, 186–87
lactation, 15, 56, 57, 59, 60, 61, 139, 140
lameness, 64, 114, 133, 139
Lasalocid, 102, 144
lathyrism, 79
leptospirosis, 65
levamisole, 65, 197
Levasol, 197
lice, 30, 136–37
ligaments, in labor, 68
linebreeding, 49
listeriosis, 65, 138, 150, 160, 231
liver fluke, 30, 144

lockjaw, 163
lungworms, 142, 144
　treating, 148
Lutalyse, 59, 182, 184, 212

M
Malta fever, 106
massage, 217–19
mastitis, 108, 113, 139–41, 202, 206
memorial, 227
meningeal worm, 142–44
meningitis, 140
Micotil, 177
microchipping, 97–100
microscope, 33
milk,
　fever, 22, 60, 201
　goiter, 10
　gruel, 207
　increasing supply, 205
　of magnesia, 8, 191
　stand, 16
minerals, 15, 20, 21, 22, 23, 24–45, 60, 178–79, 226
mint, 209
monensin, 103, 198
molasses, 83, 94, 128, 166, 200, 205
morantel tartrate, 124, 198
moxidectin, 31, 197
mullein, 216
Mu-Se, 179
mycoplasma, 139, 140, 145, 147, 172, 176, 194, 202, 232
　polyarthritis, 176

N
navel ill, 73, 81, 87
Naxcel, 17, 43, 147, 173, 174
necropsy, 14, 225
neomycin, 189, 194
neosporin, 194
nitrates, 66, 152, 153
Nolvasan, 194

Normosol-R, 187
Nuflor, 129, 130, 174–75
Nutridrench, 57, 83
nutrition, 51–66

O
omasum, 7–8
oregano, 207
　oil, 216
orf, 161–62
ovine progressive pneumonia virus (OPPV), 107–12
oxalate, 26, 164
oxfendazole, 48, 79, 107, 197
oxytetracyline, 116, 125, 145, 146, 172–73, 202
oxytocin, 73, 183–84

P
pain, 4, 11, 88, 92, 96, 97, 139, 140, 181, 192
　abdominal, 105, 131
　reliever, 181, 202, 204
pasture, 18, 20, 30, 152
parasites, 19, 22, 30–31, 32, 33–36, 44, 101, 134, 142–44, 207, 215, 216, 229
paratuberculosis, 134
penicillin, 132, 133, 138, 173, 174, 195, 196, 220, 231
phosphorus, 22, 23, 26, 54, 57, 58, 60, 164, 165
pinkeye, 64, 115, 120, 145–46
　treating, 172, 194, 202, 207
placenta, 66–67, 70, 72, 73, 75, 76, 77, 156, 160
　eating, 73
　retained, 22, 73, 106, 207
pneumonia, 64, 80, 108, 115, 127, 129, 140, 147–48, 169, 172, 174, 188–89, 220
　aspiration, 55, 72, 83–84, 171, 203
　in kids, 148
　interstitial, 107, 108
　mycoplasma, 140, 220
　ovine progressive, 107, 147
poisoning, 17, 20, 21, 63, 65–66, 79, 131, 160, 152–53, 154–55, 203–04, 224, 232
　blood, 87

poisonous plants, 15, 152, 232
polioencephalomalacia, 24, 83, 104, 149–51, 178, 182, 205, 231
port wine, 146, 202
potassium, 22, 26, 54, 85, 166, 186, 190
pre-breeding, 47, 109
pre-kidding, 213
Predef, 183
probiotic, 191
production, milk, 49, 56, 65, 73, 106, 108, 113–14, 116, 117, 132, 139, 207, 210
prostaglandin, 181
prussic acid, 154–55
pseudotuberculosis, 113
pulse,
 normal, 2, 130
 rapid, 155
pyrethrins, 137

Q

Q Fever, 64, 156
Quality of Life Scale, 221–223
Quest, 197

R

rabies, 42, 157–58, 160, 233
Rainbow Bridge, 227
raspberry leaves, 206, 213
records, 50, 66
respirations, normal, 2
retained placenta, 106, 207
reticulum, 7–8
Respirot, 185
rhododendron, 15, 66
ringwomb, 72
ringworm, 159, 204
Rompun, 89, 90, 185
routes of medication administration, 171
rumen, 4, 7–8, 17, 24, 26, 105, 123, 154, 149, 150, 151, 161, 169, 188, 191
Rumensin, 198
ruminations, normal, 2

S

SCSRPC, 30–31
SRLV, 107, 108
Safeguard, 31, 197
salal, 55, 204, 220
salmonellosis, 65, 132
scours, (see also diarrhea) 83, 188, 189, 204
scrapie, 160, 233
scurs, 92, 193
SCC, 141
selenium, 22, 23, 24, 47, 52, 178–79, 226
 deficiency, 80
 poisoning, 66
semen, 48, 51, 52, 109
shelter, 1, 13, 43, 44, 63
shooting, 223
skunk cabbage poisoning, 79
slippery elm, 205, 206
somatic cell count, 139, 141
soremouth, 161–62, 204
stanchion, 16, 27, 99
staphylococcus, 113, 120, 139, 161, 172, 184, 195, 201
star-gazing, 150
subnormal temperature, 51
sulfadimethoxine, 103, 188, 189
sulfaquinoxaline, 103, 188, 189
sulfur, 22, 26, 118, 119, 149, 151
Sulmet, 103–104
sunburn, 162
supplement, 21–26
Synanthic, 79, 197

T

tansy ragwort, 18, 66
tapeworm, 142–43
tattooing, 97, 99–100
teeth, 90, 170, 210–11
 grinding, 4, 69, 105, 140, 181, 230, 231
temperature,
 elevated, 5, 129–30, 131, 140, 147, 170
 low, 53, 181
 normal, 2, 43, 150, 181
terramycin, 145, 146, 194

tetanus, 163
    antitoxin, 90, 94, 131, 132, 163
tethering, 13
tetracycline, 116, 133, 156, 173
thiamine, 24, 104, 132, 138, 149–51, 178, 182, 198
tick bite, 65, 156
tilmicosin, 177
toxemia, 53, 58, 59, 162
toxoplasmosis, 47, 65, 80
trimethoprim, 188
tube feeding, 82, 83–84, 229
Tylan, 148, 175–76
tylosin, 175
Tyzzer's disease, 124–26

U

udder problems, 5, 108, 120, 139–41, 147, 162, 204, 206
    in pregnancy, 69
    infusions, 195
urinary calculi, 22, 43, 60, 94, 164–66
urine scald, 48, 194, 207

V

vaccination, 37, 42, 47, 48, 52, 114, 163, 157–58, 234
vaccine, 38, 42, 47, 64, 106, 110, 114, 115, 117, 147, 162, 163, 167, 171, 79, 231
vacutainer, 40–41, 238
Valbazen, 48, 31, 79, 128, 143–44, 148, 197
valerian, 94, 206
Velenium, 179
vinegar, apple cider, 26, 207
vitamins, 177

W

WADDL, 36, 40, 110, 231, 236
weight loss, 26, 106, 107, 108, 135–35
wether, 14, 22, 43, 93, 96, 164–66, 194, 228
white muscle disease, 22, 85
wormwood, 207, 209, 215

X

xylazine, 89, 90, 185

Y

yersiniosis, 65

Z

zinc, 22, 23, 24, 118, 119, 121
    oxide, 48, 194
    sulfate, 133